Apostle of Progress

THE MEXICAN EXPERIENCE
William H. Beezley, series editor

Apostle of Progress

Modesto C. Rolland, Global Progressivism, and the Engineering of Revolutionary Mexico

J. JUSTIN CASTRO

University of Nebraska Press
LINCOLN & LONDON

© 2019 by the Board of Regents of the University of Nebraska

Segments of this book relating to Baja California were published in "Modesto C. Rolland and the Development of Baja California," in *Journal of the Southwest* 58, no. 2 (Summer 2016): 27–46. A previous draft of chapter 8, "A Stadium for Stridentopolis," was published in Spanish as "Un estadio para Estridentópolis: Modesto C. Rolland y su visión moderna de Xalapa," *Balajú: Revista de Cultura Comunicación de la Universidad Veracruzana* 3, no. 5 (August–December 2016): 3–17.

All rights reserved
Manufactured in the United States of America

Library of Congress Cataloging-in-Publication Data
Names: Castro, J. Justin, 1981– author.
Title: Apostle of progress: Modesto C. Rolland, global progressivism, and the engineering of revolutionary Mexico / J. Justin Castro.
Description: Lincoln: University of Nebraska Press, [2019] | Includes bibliographical references and index.
Identifiers: LCCN 2018009208
ISBN 9781496211736 (cloth: alk. paper)
ISBN 9781496211743 (pbk.: alk. paper)
ISBN 9781496212498 (epub)
ISBN 9781496212504 (mobi)
ISBN 9781496212511 (pdf)
Subjects: LCSH: Rolland, M. C. (Modesto C.) | Civil engineers—Mexico—Biography. | Mexico—History—20th century.
Classification: LCC TA140.R83 C37 2019 | DDC 624.092 [B]—dc23 LC record available at https://lccn.loc.gov/2018009208

Set in Vesper by Mikala R. Kolander.

Contents

List of Illustrations vii
Acknowledgments xi
Introduction: Matters of Perspective xiii

1. Child of the Porfiriato, Child of the Periphery 1
2. The Reluctant Revolutionary 21
3. A Mexican Progressive 39
4. Back to the Periphery 61
5. War and Peace 82
6. Transitions 103
7. Opportunity, Defeat, and the Death of Virginia Garza de Rolland 120
8. A Stadium for Stridentopolis 140
9. Mr. Bothersome 158
10. The Undersecretary 177
11. Going Big 196
12. Out of the Ports and into the Hills 216
Conclusion: Final Thoughts about Modesto Rolland's Life and Legacy 235

Notes 247
Bibliography 279
Index 301

Illustrations

Following page 102

1. Students at the Colegio Rosales, 1898
2. Mexico City, ca. 1890
3. Rolland as a young engineering student, 1905
4. The de la Garza family, 1908
5. Portrait of de la Garza during her engagement to Rolland, 1907
6. Workers building the Xochimilco–Mexico City aqueduct, 1908
7. New members of Porfirio Díaz's last cabinet, including Rolland's mentor, Manuel Marroquín y Rivera
8. Ad for Rolland's concrete workshop, ca. 1911
9. Blueprint for Rolland's patent filing for a reinforced-concrete water tank, 1913
10. Diagram of a reinforced-concrete process Rolland patented, 1913
11. The Rolland family, 1913
12. Street scene during the Ten Tragic Days in Mexico City, 1913
13. Military cadets at Chapultepec Castle, 1913
14. U.S. soldiers raising the American flag during their occupation of Veracruz, 1914
15. Venustiano Carranza's cabinet members and close advisors, 1915

16. Rolland with Gen. Álvaro Obregón, ca. 1915

17. Henry George, ca. 1880

18. U.S. progressives Crystal Eastman and Amos Pinchot, ca. 1915

19. U.S. progressive Lincoln Steffens, 1914

20. Illustration representing U.S. frustration over the Punitive Expedition, *New York Herald*, 1916

21. Illustration by Nelson Harding, *Brooklyn Eagle* and *Current Opinion*, 1916

22. Adolfo de la Huerta, interim president of Mexico, 1920

23. Obregón around the time he assumed the presidency, ca. 1920

24. The El Buen Tono booth and Modesto C. Rolland at the Mexico City Grand Radio Fair, 1923

25. Ad for the free ports in the Mexico City newspaper *El Demócrata*, 1924

Following page 176

26. Bust of Heriberto Jara outside Xalapa Stadium

27. Construction of Xalapa Stadium, 1925

28. A drawing from Modesto Rolland's plans for the Xalapa garden city, 1925

29. Wedding photo of Rolland and Rosario Tolentino with the Tolentino family, 1926

30. Tolentino, Rolland, and Rolland's children, 1927

31. Tolentino and her child with Rolland, Ana María, 1929

32. Hotel Chula Vista, ca. 1930

33. Pamphlet for Rolland's Aero-Motor México water pump, 1932

34. Tolentino and Rolland, ca. 1935

35. Tolentino, Ana María, and Rolland in Xochimilco, 1938

36. Illustration of James Eads's ship railway in *Scientific American*, 1880

37. Illustration of the ship railway designed by Rolland, 1946

38. Artist's rendition of the ship railway concept, 1949

39. View of Córdoba, Veracruz, from Rolland and Tolentino's Rancho Santa Margarita, 1941

40. Tolentino, Ana María, Rolland, Martha, and Catherine Rolland, Rancho Santa Margarita, 1941

41. Rolland's grandchildren playing at Rancho Santa Margarita, 1947

42. Pres. Manuel Ávila Camacho and Rolland discussing the City of Sports, 1944

43. Officials visiting the construction site for the City of Sports, 1944

44. Construction of the Plaza de Toros, ca. 1944

45. Aerial view of the City of Sports under construction, ca. 1945

46. The fixed dredge, from the cover of the free ports booklet *Draga fija*, 1950

47. Rolland with workers and a foreign specialist at Salina Cruz, ca. 1950

48. Rolland family photo, Mexico City, 1952

49. Rolland family photo, Córdoba, 1959

50. Rolland, Tolentino, and family celebrating Rolland's eightieth birthday, 1961

Acknowledgments

Rob Alegre, Airek Beauchamp, Enrique Bernales, Jürgen Buchenau, Kellie Buford, Jacob Canton, Barry Carr, Angela Castro, Joseph Dale Castro, Olivia Castro, Kevin Chrisman, Ryken Cocherell, Geoffrey Clegg, David Dalton, Paul Eiss, Omar Escamilla, Ben Fallaw, Raphie Folsom, James Garza, Karl Jacoby, Joe Key, Chris La Puma, Carrie Larson, Michael Matthews, Gary Moreno, Vicent Moreno, Andrew Paxman, Jayson Porter, Elissa Rashkin, Miles Rodriguez, Jorge M. Rolland C., Manuel Arturo Román Kalisch, Terry Rugeley, Aaron Russell, Juan José Saldaña, Marcela Saldaña, Laura Surdyk, Emily Wendell, Deanna Wicks, John Womack. *Thank you.*

The completion of this book was made possible by a faculty research award provided by Arkansas State University. *Thank you.*

Introduction

Matters of Perspective

DURING ONE OF MY research trips for this book I spent a month driving across the Baja California peninsula. I revisited John Steinbeck's *The Log from the Sea of Cortez* and soaked in *The Pearl*. Growing up in the United States back and forth between California and Oklahoma in a working-class family, I've always related to Steinbeck, who, in addition to writing stories about migrating Okies and down-and-out California laborers, wrote about Mexico, another facet of his life that I relate to. Steinbeck excelled at verbally painting images of people and places. He spoke his truth loudly, clearly, and with the choicest words. But as brilliant as Steinbeck could be, his books, particularly *The Log from the Sea of Cortez* and *The Pearl*, have a sort of cinematic tendency to pit good versus evil, rich against poor, and a romantic past against a brutal present. They were written in terms of stark contrast, with clear winners and losers. The research I conducted for my own book has led me to messier conclusions. What gets lost in some of Steinbeck's stories are the morally and politically ambiguous characters, characters like most of us, and in this case like the person whose life I've spent the last few years trying to reconstruct: Modesto C. Rolland, a man little known today but who was in his time one of the most talented civil engineers in Latin America, a man who played no small part in building modern Mexico.

Steinbeck's *The Pearl* takes place in La Paz, which happens to be the town where Rolland was born in 1881. The novella is a story of economic change and the greed that came to the region via an expanding pearl economy and white foreign business operators. In the story, set in the early to mid-twentieth century, a scorpion stings a poor young Indian couple's baby, named Coyotito, leaving them desperate to find help. A selfish doctor refuses to aid them. That is, he refuses to assist until the father, Kino, finds "the Pearl of the World." The rest of the tale is a snowballing tragedy in which greedy men try to rip off Kino and his wife, Juana. Fleeing La Paz north to escape attackers and to obtain a better price for the pearl, Kino and Juana run into trouble with trackers hot on their trail. These human bloodhounds mistake Coyotito's cries for those of a coyote and accidentally shoot the child in the head, killing him. Kino kills the trackers, and then he and Juana return with their child's dead body to La Paz and throw the cursed pearl back into the ocean.

Steinbeck detested what he perceived as a greedy and racist modern world, as well as how that greed and racism tainted ancient Baja California societies, leaving behind its indigenous inhabitants, wrecked in the wake of change.[1] He was not against all notions of progress, however. He firmly believed in education and the benefits of cutting-edge medicine. He had a distaste for what he saw as ignorant superstition.[2] But he thought that his world often ignored or took advantage of poor, rural communities. Steinbeck was certainly correct about forces outside of La Paz bringing drastic change and the avarice excited by the pearl industry. The modern world has often been hard on small communities, but Steinbeck's view of the Baja California peninsula was limited by his biases and sometimes by his self-righteous activism. Many people in La Paz embraced modernity, including young indigenous men who dove for pearls, mestizo families who intermarried with foreign adventurers, and business operators. The town would grow into a thriving and picturesque city, stealing people from other towns and countries. The legacy of La Paz residents and those visitors who have interacted with them has been much more mixed than Steinbeck's tragic morality tale. Some children would benefit from incoming foreigners and connections to the greater world, unlike poor Coyotito. Rolland was one of those children who benefited; he grew up in different circumstances. While Steinbeck romanticized the primitive wonder of poor Baja California villagers, the adult Rolland argued that Baja California had to embrace the modern world and that doing so would be beneficial to its people. His record as a prophet proved to be hit-and-miss, but he, far more than Steinbeck, tried to rethink the position of Baja California, and indeed of all of Mexico, in the emerging twentieth-century world of technological advancement and globalization.

The history told in the following pages is not about Steinbeck or the pearl industry, though I do start the first chapter with a story about a pearl. Nor does this book focus solely on Baja California. This history is about the messiness and complexity that were so often part of the Mexican Revolution, the construction of modern Mexico, and the international spread of ideas and technologies. I start this book with a nod to Steinbeck not only because I appreciate his passionate writing but also because *The Log from the Sea of Cortez* and *The Pearl*, which are still widely read in the United States, represent common and unsophisticated tropes about Mexico. Steinbeck was not alone in presenting a romantic view of an ancient Mexico filled with primitive wisdom. Most people I engage with on a daily basis, Americans—in the "from the United States" version of the word—see Mexico as a land of beaches, tacos, violence, wild adventures, and a splash of mysticism. One goal I have for this book is to complicate that image. My story is fundamentally different. It is a story about a professionally ambitious middle-class engineer who loved his nation and wanted to improve it by adapting profound liberal ideas to complex Mexican realities. It is also an exploration and critique of engineers, technocrats, progress, development schemes, and the modern world. I tell this history through the lens of Modesto Rolland, an immensely talented if not always likable engineer.

Rolland's life provides an exciting narrative in which to explore Mexico's tumultuous history during the late nineteenth century and the first half of the twentieth century. He obtained his education and began his career as a young teacher and engineer during the long dictatorship of Porfirio Díaz Mori (1876–1880, 1884–1911), a period of modernization. Rolland subsequently became one of the backers of Francisco I. Madero, a failed presidential-candidate-turned-revolutionary who served as president from 1911 to 1913. At the time of that transition, Rolland was already a civil engineer and one of the foremost experimenters with reinforced concrete. Following the assassination of Madero in 1913, Rolland became a communications official for the Constitutionalists—the rebel coalition that picked up the banner of revolution after Madero's death—and their leader, Venustiano Carranza. Rolland served as an important researcher and propagandist in the United States for the Constitutionalists. From there Rolland helped carry out agrarian reform in Mexico, explored for petroleum, built massive infrastructure projects, developed ports, served as a high-ranking bureaucrat in different government ministries, built some of Mexico's most impressive stadiums, drew up plans to carry ships across Mexico on trains, and wrote essays and policies about municipal governance, trade, and taxation. And he did all this while successfully navigating a chaotic era of violent turmoil. Rolland did not retire completely from government service until the 1950s. In other words, he had his hands in many of the events, trends, and developments that created modern Mexico. Yet, few people have heard of him.

In this telling of Rolland's life and times I make the case that Rolland represents an often overlooked technocratic, moderate ideology shaped by nineteenth- and early twentieth-century liberalism, the rise of the social sciences, and a global progressive movement that thrived during the late Porfirian period (1890–1911) and the revolutionary era (1911–46). Mexican intellectuals widely debated progressive ideas, but such ideas were most pronounced in those agents who, like Rolland, spent considerable time abroad. Upon their return to Mexico, these individuals influenced Mexican policies and development schemes made and undertaken by the governments that rose out of the revolutionary fervor. Not all of Rolland's technocratic compatriots shared his exact vision, but most of them engaged with the debates common to progressive movements under way in the United States, Europe, Australia, New Zealand, and parts of South America. Engineers and other technocrats incorporated these ideas into infrastructural, economic, scientific, journalistic, and legislative pursuits.

The variation among these figures was not only a product of the diversified concepts and people associated with progressivism; as the editors of the influential collection of essays *The Social Construction of Technological Systems* state succinctly, "System builders are no respecters of knowledge categories or professional boundaries."[3] And that is exactly what many of these revolutionary technocrats were, or at least aspired to be: system builders—people who meshed intellectual trends and new technologies to remedy long-standing problems through social, material, and infrastructural development during a period of revolutionary change. Progressivism was an intellectual river made up of a number of inflowing streams. That the movement was ideologically varied and malleable is what made it appealing to system builders.

Presidents and other top officials found that they often had to contend with these technocrats and their ideologies, because, like Rolland, they belonged to a small class of Mexican nationals who could reliably develop serious economic and infrastructural plans and turn them into physical reality. This is one reason Rolland thrived professionally for so long, in a violent period, and across so many government administrations. He was not particularly likable, but neither was he easily replaceable. Of course Rolland had to contend with presidents and other political leaders, men who were only sometimes motivated by the same ideals that drove him. When he and his superiors failed to come to terms, projects fell apart; progress, something Rolland genuinely believed in, stalled.

Another reason for Rolland's longevity was his ability to take advantage of the often contentious political differences between bureaucracies, governors, and presidents in revolution-era Mexico. Rolland partnered with Salvador Alvarado, the military governor of the southern Mexican state of Yucatán from 1915 to 1918, in an attempt to carry out agrarian reform policies that were more radical than those promoted by "First Chief"-turned-President Carranza (1913–20). When Pres. Plutarco Elías Calles (1924–28) ousted Rolland from the federal government, Rolland found refuge and opportunity in the state of Veracruz, where his then ally, Heriberto Jara Corona, was governor. When Rolland had a falling out with one of the secretaries of communications and public works who served in Lázaro Cárdenas's government (1934–40), Rolland found a friendlier home in the Secretaría de Economía Nacional, or Ministry of the National Economy. Rolland often referred to himself as apolitical, and he does appear in some ways to have disliked politics, but he was politically astute.

This self-proclaimed apoliticism is another point worthy of deeper exploration. Rolland was not alone in professing his political neutrality. Many of his engineering peers also painted themselves as nonpartisan. They argued that their advice was unbiased scientific thought. In reality, however, little escaped politics. Rolland and others like him always took sides. Their largest projects needed approval from political leaders. And, as the historian Paul R. Josephson argues in his brutal critique of large-scale engineering projects, it has been impossible to divorce "the facts of how nature operates from the political decisions [required] to transform nature for the betterment [or detriment] of humankind."[4] But there were reasons for Rolland's position. It was a stance taken by a number of "scientific" city planners, sociologists, agronomists, and engineers in the progressive world. He took up this apolitical label to shield himself from near-constant political turnover and rebellions. Rolland and a number (though not all) of his engineering peers were genuinely perturbed by clientelist and populist politics. For its own betterment, they thought, Mexico needed consistent policies supporting development that would be free of political pandering and power plays. The fact that Rolland became the head of many prominent "apolitical" think tanks shows that he was not alone in this line of thought; others respected his view. Being apolitical was also a way for Rolland to state that he himself was not a violent contender for power, signaling that political and military leaders could trust him not to plot against them. His goal was to build things based on empirical studies; politics were secondary.

Rolland's claim of complete objectivity is more than deserving of the utmost criticism—as will become clear, he became a rather successful propagandist—but instability and abrupt policy changes based on personal favoritism and political expediency were real problems for Rolland and the young revolutionary state. Engineers and other technocrats gathered together under the banner of apoliticism in an attempt to provide a stable path in a world of political and military turmoil. As politicians and military leaders duked it out, Rolland saw engineers as building a more permanent revolution within the material structures of Mexico—the city streets, highways, railroads, ports, irrigation, housing, factories, and stadiums. To him and many other engineers, the built environment was ultimately most important for the creation of long-lasting peace and prosperity. Their impact was substantial and remains visible, but engineers encountered immense difficulties, never escaped politics, and failed in their quests as often if not more than they succeeded.

How Rolland went about designing and building that environment tells a lot about larger trends that influenced him and others. Embedded in his writings and projects are clues to the evolution of intellectual streams and engineering trends. Rolland's undertakings incorporated a potpourri of practices taken from his Mexican predecessors, his peers, and foreigners alike. He borrowed concepts from nineteenth-century liberalism and positivism while interacting extensively with new progressive and modernist trends. This intermixture can be seen in the social goals of his designs. For example, Rolland consistently pushed for state-directed infrastructural projects that increased national solidarity and the power of an increasingly technocratic central government, a Mexican counterpart of sorts to the "wartime socialism" seen in the United States and Europe during World War I.[5] But his vision of centralization initially allowed for semiautonomous regional development. He wanted government protections and federal infrastructure, but he also pushed for the central government to allow for freer elections and more independence in local planning, at least with regard to public utilities and a certain tax policy that became an obsession for him. Rolland argued that government aid to regional projects and better infrastructure linking regions and Mexico City provided a more democratic and sustainable means of development. It was an argument based on municipal government reforms under way in the progressive world but also a legacy of nineteenth-century Mexican federalism. It is nevertheless clear that he thought the power of the central government was crucial to his larger projects and to fixing historical wrongs and expanding social justice. It was a difficult web to navigate. Because of the complexities involved, Rolland's ideas come off at times as paradoxical. His life presents a strong example of the contradictions and changes taking place in the transformation of nineteenth-century Mexican liberals and *científicos* into mid-twentieth-century technocrats.

I hope that my academic peers find *Apostle of Progress* to be a compelling work of scholarship, but I wrote this book with a broader audience in mind. I wanted to create a narrative story that provides students and interested members of the general public an enjoyable and engaging entry point into the dynamics of twentieth-century Mexican history. With this in mind, I am going to make my historiographical discussion brief.

This book draws from a number of historiographical branches. Influences include works in biography and narrative history, the history of progressivism, the history of science and technology, and the history of the Mexican Revolution and its legacies. For students and general readers interested in delving deeper into these scholarly threads and other topics covered in this book, a close reading of my notes and bibliography will provide a start down those paths.⁶

In my exploration of progressivism I build on publications written by a number of outstanding historians who have worked on revealing the international breadth of progressive thought, policies, and social action in the late nineteenth and early twentieth centuries. The historian James T. Kloppenberg's explanation of how U.S. progressives changed their conceptions of liberalism applies well to the thinking of many of their contemporaries in Mexico who became involved in the Mexican Revolution: "these thinkers turned the old liberalism into a new liberalism, a moral and political argument for the welfare state based on a conception of the individual as a social being whose values are shaped by personal choices and cultural conditions."⁷ Like Kloppenberg, most authors who discuss progressivism in a global context have focused on the connections between western Europe, the United States, Australia, and New Zealand. One of the firm conclusions I came to while writing *Apostle of Progress* is that this discussion of global progressivism needs to be expanded further into Mexico if we are to more fully comprehend the revolution, the initiatives that it birthed, and a more complete understanding of the world during this era. I think scholars examining other parts of Latin America would also benefit from more deeply exploring the influence of this complicated intellectual and social phenomenon.⁸

Historians besides myself have made the connection between agents of the Mexican Revolution and a more encompassing worldwide movement, though they have rarely addressed such a connection in clear terms or focused on the topic specifically. Most of these scholars have linked Mexican intellectuals and U.S. progressives without discussing Mexico as part of a larger, more global progressive phenomenon.⁹ My book fleshes out these transnational connections.

In connection with my argument on progressivism, I demonstrate that Rolland was a person essential to Carranza's savvy U.S. foreign policy. Building on the works of historians who have explored Mexican operatives in the United States during the Mexican Revolution, I argue that Carranza's agents were proactive and often successful at manipulating the U.S. public, especially through their collaboration with U.S. progressives and their media outlets. Rolland was an instrumental player in using progressive networks to solidify support for Carranza's faction.[10]

Other related gaps in the historiography I hope to address are the limited coverage of moderate, middle-class specialists and the importance of infrastructure projects to nation-state building in Mexico during the revolution and into the mid-twentieth century. Scholars—especially those who have focused on the environment, economists, agronomists and surveyors, development, and the history of technology—have begun to address this lacuna, but there is still much work to be done. I don't claim to fill this gap completely, but I think this book makes a contribution.[11]

It is also my desire to bring further nuance to studies of the Mexican Revolution and its consequences. Even if less black and white than Steinbeck, some of the most sophisticated histories on the revolution still possess a sort of polarizing, cinematic quality, with larger-than-life personalities clashing in an epic contest for Mexico's future. But the near constant focus on stark oppositions between men like the firm patrician Venustiano Carranza and the peasants' hero Emiliano Zapata has long obscured the importance of Rolland and others like him.[12] Rolland was a major player in developmental projects in Mexico from 1906 to 1952, and he was not alone. Recounting the story of the revolution without engineers and other mid-level planners makes for a very incomplete understanding of the revolution and the construction of modern Mexico. I hope to complicate these dichotomous portrayals by emphasizing a middle-out sort of history that argues that mid-level technocrats, intellectuals, and bureaucrats had a longer-lasting, more intricate, and more important impact on the conduct and legacies of the revolution than is usually acknowledged.[13]

Exploring Rolland's life and work provides a window into his technocratic worldview and, to some extent, the mentalities of many of his colleagues, people essential to the construction of modern Mexico and to what Rolland perceived as progress. Working in so many facets of development during the first half of the twentieth century, Rolland serves as a lens through which we may examine the Mexican Revolution, relations of power, engineering practices, infrastructural development, political ecology, transnational exchanges, and Mexican politics. This book is a history of one man's life, but it is also a history of engineers and the neglected personalities who during the Mexican Revolution worked between military chieftains on one side and impoverished soldiers on the other—those individuals who drew up the blueprints, printed newspapers, implemented reforms, and constructed complexity, people all too often forgotten but who built modern Mexico and created a more global world. This is a history about dreamers of progress and the doubts they created.

1

Child of the Porfiriato, Child of the Periphery

MODESTO ROLLAND WAS STILL a baby when a man who was fishing happened upon the largest pearl that many of La Paz's residents had ever seen. The town buzzed with hungry gossip anytime a diver made a valuable find, and this one beat them all. The man, according to rumor, sold the pearl for $14,000. Some said it was worth more.[1]

Pearls had long been prized in La Paz. The community was small but vibrant, nestled against the Gulf of California near the southern end of the Baja California peninsula. The region's original inhabitants—the Cora, Guaycura, and Aripe—had adorned themselves in pearls before the Spanish conquistador Hernán Cortés arrived in 1535. The Spaniards had put significant energy into acquiring these pearls. It was usually the locals who fished for a living, often young indigenous boys, who acquired the pearls. Some of the divers could hold their breath for more than two minutes while they combed the sea floor. Much later, in the 1870s, well after Mexican independence, Europeans and Americans had come with ships carrying compressed-air diving suits, pumps, and hoses. Their divers could not hold their breath as long as local boys, but encased in their technology the foreign divers could go deeper and stay under for much longer. They reaped massive profits, and many of the locals resented them.[2]

By Modesto's second birthday, in 1883, local business operators had petitioned the Mexican president, Gen. Manuel González Flores (1880–84), for extensive rights to farming the pearl-producing oysters. But the president who preceded and followed González, the far better known Gen. Porfirio Díaz, gave near-exclusive harvesting rights to five foreign companies the following year.³

While pearls found their way to Europe, Europeans found their way to La Paz. Twenty-eight years before Modesto's birth, his father, Jean François Rolland, or Juan Francisco Rolland, as he was known in Mexico, had arrived in 1853. He had left France in 1837 as a young man.⁴ He appears to have sailed the high seas before spending two years in gold rush–era California and then making his way south to the Baja peninsula, first to palm-filled Mulegé and then to La Paz.⁵ In La Paz, Juan Francisco worked as a carpenter, associating with other European immigrants. He wooed a young woman, María Jesús Mejía Altamirano, and married her in 1855. They immediately set to having a large family, eleven children in total. Modesto was the youngest.

The history of Modesto's youth and young adulthood is a story of origins and context. It allows us to explore the region from which he first viewed the world; it delves into his personal life and education as he transitioned from an ambitious student into an upcoming civil engineer–turned-revolutionary. His childhood in the Baja California peninsula, a land peripheral to the heartland of Mexico yet strangely global, left a lasting imprint on Rolland. His education in Mexico City introduced him to engineering, his wife, vibrant and contested ideas, and a turbulent political arena that ultimately brought revolution to Mexico.

The Baja California Peninsula

Most political leaders in Mexico City considered Baja California a hinterland, a peripheral frontier that was in need of incorporation into the nation-state. There was some truth to this perception. Not all of the area was (or is) hospitable to human settlement; there were few roads, and the entire peninsula was thinly populated. Just decades before Modesto's birth U.S. soldiers and settlers had taken much of Mexico's northern territory. But Baja California was not as desolate as some imagined. The modern world was not absent in Baja California, nor were its residents passive observers. Mexico City's grip on the region was weak but growing, while the peninsula's location on the Pacific coast just below the vibrant and aggressively expanding United States provided access to people, goods, and ideas from around the world. La Paz had seen visitors from far-flung lands come and go for centuries. In this sense it was a prime place to raise an apostle of progress like Rolland.

La Paz, at the time Modesto was born and baptized in 1881, was interconnected with global trade, but it was by no means a huge place. It was home to perhaps six thousand people.[6] Adults and children alike combed the clear ocean water for pearls, shells, and a variety of sea life. Grape growers just outside of town produced wine for Catholic liturgies and for small stores. Local carriages and pedestrians passed fish and fruit vendors displaying the day's harvest. Beyond La Paz, mining had grown substantially, becoming a valuable if destructive industry. Newcomers from around the world began to come to the region in search of copper and gold as much as pearls. But with the exception of the nearby agricultural lands, small ranches, and mines, La Paz remained isolated except for ship traffic; it was a coastal oasis surrounded by sea, mountains, and desert. Still, Europeans, Americans, Asians, and Mexicans from across the gulf stopped by the port, bringing a metropolitan air to the town.[7]

Other places in the world were changing rapidly. The nineteenth century had ushered in an era of scientific discoveries and technological innovation. European factories produced locomotives, telegraphs, machine guns, and cheap manufactured goods. Sometimes sailors brought new curiosities with them, such as the compressed-air diving suits.

The machines in turn changed how many white Europeans and Americans from the industrializing north viewed other people, including indigenous and mestizo Mexicans. The British and their descendants in the United States began to see, as one historian put it, "machines as the measure of men," equating technological development with societal worth.[8] Machines such as trains displayed sophistication, metallurgical skills, and, just as important, ambition, rationality, and timely precision.

Forgetting their own economic "backwardness" just centuries before, many Europeans believed their devices, and their ability to use such technology to exploit colonies, were the products of an enlightened, industrious, and inherently better people.

These beliefs were not limited to the halls of Britain, Germany, and the much closer United States; political leaders and intellectuals in Mexico City who surrounded Presidents González and Díaz possessed a similar view of the world. They too desired their country to be modern, industrialized, and strong. They wanted infrastructure, machines, and the wealth that came with them. And they feared that if they did not catch up with western Europe, their country would languish as a land filled with Indians who feigned allegiance, produced little for the growing capitalist market, spoke different languages, and wallowed in lethargy. In the eyes of these politicians and their technocratic advisors, places like the Baja California peninsula possessed a rough beauty but remained remote, underpopulated, and undeveloped—exactly the sort of place where premodern barbarism could still flourish.

The Díaz administration did much to modernize urban Mexico and to increase state centralization. Under the motto of "order and progress," Díaz's advisors, many of whom were referred to as *científicos* because of their belief in "scientific" prescriptions for society and politics, built an environment enticing to foreign capitalists on whom the Díaz administration relied to build railways, telegraph stations, and harbors. In exchange, foreign businesses gained easier access to natural resources. By the 1870s telegraphs had connected the Atlantic and Pacific coasts like a collective extension of the nervous system.[9] By the end of the next decade railroads were crisscrossing the nation, though often the trains were carrying raw materials to the United States. The increase in communications helped facilitate foreign and domestic trade and the government's ability to quash rebellions incited by people who did not share Díaz's vision.

In addition to infrastructure, Díaz invested heavily in education. Justo Sierra, as minister of education, promoted public schooling as the best means to create a unified Mexico. He believed that indigenous and mestizo peoples, many of whom cared little about the concept of Mexico, needed to learn to be Mexican and in the process to learn to be modern. To create Mexicans, the president and the Congress extended free and obligatory public education to the whole of the nation in 1888. The reality fell far short of the ideal of universal education, but school attendance increased significantly. From 1877 to 1910 the percentage of funding for education at the state and federal levels more than doubled. Díaz and Sierra also opened a number of colleges.[10]

Modesto's family also held education in high esteem. His mother's family had a history of producing teachers, and his sisters continued that tradition. They taught at schools in Mulegé, La Paz, and in a new copper-mining town by the name of Santa Rosalía. Rolland's mother constantly pushed him to excel in school, and she remained a motivating force during his youth and young adulthood.[11]

Juan Francisco, María, and Modesto moved to Santa Rosalía sometime around 1886. Modesto was a small child, but most of his brothers and sisters had already established homes of their own, mostly in and around La Paz. Three hundred miles to the north Santa Rosalía stood uncomfortably pressed between desert hills and the Sea of Cortez. A French mining company, El Boleo, founded the town in 1885, though miners had been attracted to the rich mineral deposits in the area since the 1860s, perhaps earlier. Backed by money from the Rothschilds, a European Jewish family that had amassed one of the greatest fortunes in the world, El Boleo took advantage of mining laws that provided private interests with access to subsoil rights.[12] The Rolland family was one of approximately seventy families that the company persuaded to build the townsite. They stayed until 1890.[13]

The Díaz administration meanwhile moved to further consolidate its control over the peninsula. In December 1887 Díaz reorganized the Territory of Baja California and its three components—south, central, and north—into two districts:

the south, ruled from La Paz, and the north, governed from Ensenada. To the distaste of many locals, Díaz placed non-Bajacalifornianos into top political posts. This was not a new practice. But despite the attempts to draw the peninsula further into the federal system, local identities still outweighed national allegiance. The federal presence was nonetheless growing. Engineers and technicians were slowly extending telegraph lines and improving the Mexican navy's small flotilla. The capital's tentacles of progress were becoming more entrenched, almost hidden, coming through slender cables and into the fibers of clothes and the gears of engines. In addition to creating policy, agents from the nation's capital diffused technology outward as a way to extend their authority and to limit the power of foreigners.[14]

After the brief stint in Santa Rosalía the family returned to the yellow sands and plastered walls of La Paz. Modesto's mother continued to compel her youngest son to excel in school.[15] Rolland had grown taller, but he was still a child. He had deep brown, almond-shaped eyes, and Rolland's mother parted his dark cowlicked hair to the left. When not at home or in class he sometimes strolled the boardwalk.

Rolland's Education in Culiacán and Mexico City

In 1896, urged on by his mother, Rolland embarked on a journey across the gulf and into the state of Sinaloa to train to become a teacher. He was fifteen. Modesto had been admitted into the Colegio Rosales in Culiacán, likely on a small scholarship. Culiacán was no New York or Mexico City, but it had its own form of vibrancy. The large, two-towered cathedral contrasted with the green mountains just beyond the cityscape. Palm trees lined the dusty streets, where people worked and children played.

Established in 1874, Colegio Rosales was one of the regional schools grew during the Porfirian era. Although less prestigious than the older universities of Mexico City, the Sinaloan college had become an important center for learning. The evangelists of science, industry, and enlightenment had come with rulers and pencils to prepare teachers, accountants, metallurgists, agronomists, and engineers. Initially Rolland faced ridicule for being an "overgrown boy."[16] But he later reflected that his intellect had won him the respect of teachers, who began to give him private lessons. Rolland's confidence grew. As the new century dawned, Rolland wrapped up his coursework. He was excited for what felt like inevitable change.[17]

His next professional decision displayed a hunger for achievement, financial success, and acclaim. After graduating from Colegio Rosales, Rolland, almost twenty years old, set out for Mexico City.[18] He caught a train to the national capital, and during that journey he saw the interior of mainland Mexico for the first time. He had probably read all about locomotive travel, how people watched the world rush by in a blur of color and smoke. Over the tracks he saw the Sierra Madre Occidental, the valleys of central Mexico, small villages, and emerging cities.[19]

He arrived in Mexico City in January 1901. The move took guts. He had no family there, not even an acquaintance. The city had expanded during Díaz's rule to more than a half million people. Electric lines crisscrossed the streets, where trolleys carried young urbanites and sometimes ran them over. Display windows exhibited typewriters, accordion cameras, and French perfumes. Large pointed sombreros mixed with *charro* regalia and top hats. Walking into the elite watering hole known as the Jockey Club, black-suited men escorted "S-shaped" women who wore fine jewelry and corseted gowns.[20] Soon Mexico's first automobiles would sputter between the clacks of horse-drawn carriages. There were powerful smells, too, including mixtures of El Buen Tono cigarettes, roasting corn, perfume, and human waste.[21]

Rolland struggled but thrived in the capital. His mind bustled. He considered returning to school to become a medical doctor. He talked with the directors of the National School of Medicine, who told him that a medical degree would take five or six years of study and additional practice to complete. Rolland also spoke with administrators at the recently established National School of Engineering, who told him that he could complete a degree there in four years. Rolland decided to become an engineer.[22]

In addition to adjusting to the bustle of the city, Rolland also learned the challenges of navigating educational bureaucracies, an education in itself that would serve him well. To start formal training, he first had to return to Sinaloa to gain signatures from the director of Colegio Rosales and the governor of Sinaloa to formally certify his teaching certificate, which did not happen until mid-December 1903. That same year he also managed to take a calculus course at the Scientific and Literary Institute of the State of Hidalgo in Pachuca, in the mountains north of Mexico City.[23] He had wrapped up his coursework in Mexico City by 1905. Assured of his ability, professors began to more directly mentor Rolland.[24]

Rolland entered the world of Mexico City higher education at an exciting time. Scientists had been active in relatively new societies such as the Antonio Alzante Scientific Society and the National Medical Institute. In the 1890s the city had hosted the Second Panamericana Medical Congress and the Tenth International Geological Congress. Intellectuals participated in robust discussions about the newest social theories from Europe, including the positivist theories of Auguste Comte and the social Darwinism of Herbert Spencer, but also ideas of utopian socialists, including Edward Bellamy, Henri de Saint-Simon, and Charles Fourier. Students, professors, and officials debated how to incorporate these ideas—among many, many others—into efforts at tackling the difficulties faced by an expanding Mexico City. Engineering students in particular took a strong interest in city planning. They considered themselves practical scientists who would apply their studies to the everyday lives of Mexicans far more than would their peers in the humanities. Engineers would build the world that others merely discussed.[25]

The capital had a long history of engineering, especially pertaining to water and mining. The Aztecs had built aqueducts to Tenochtitlán, the predecessor of Mexico City. The Spanish colonial government began works on the *desagüe*, or draining of the ancient lakes that surrounded the city, in 1607. By the late 1600s military engineers were playing an important part in the city's drainage and water works, a role that increased during Bourbon rule in the following century.²⁶ In 1792 the colonial government established the Colegio de Minería to improve mining techniques. Since the earliest days of Spanish colonization, officials had been interested in exploiting Mexican metals, especially silver. The professional training of engineers gained more momentum after Mexican independence, during the administration of Pres. Benito Juárez (1867–72), which created the Escuela Especial de Ingenieros, or Professional School for Engineers, in 1868. As infrastructure and public works projects (and a reliance on foreign specialists) boomed during the subsequent administration, that of Porfirio Díaz, engineering officially became a national priority.²⁷ In the 1880s state leaders changed the name of the engineering school to the Escuela Nacional de Ingeniería, or National School of Engineering. They increased the number of scholarships available to train talented students, including people who became prominent in Porfirian governance and education. One such student was Manuel Marroquín y Rivera, who later, as a professor, worked on Mexico City's waterways and served as one of Rolland's close advisors.

The worldwide growth of urban studies that strove to improve society through the rational planning of parks, roads, and ports created a shift in the programs at the National School of Engineering. During the 1890s and early 1900s the school began emphasizing the production of civil engineers. An 1897 policy placed road, port, and canal studies under civil engineering; they had been separate before. Students also noticed that civil engineering had become more lucrative, as infrastructure projects increased in number.²⁸

Most of Rolland's training, as well as that of his engineering-student peers, focused heavily on mathematics and a broad education in engineering practices: topography, hydrology, and infrastructure. Their texts came from a mix of French, German, U.S., and Mexican authors. They read the works of the U.S. civil and hydraulic engineer Mansfield Merriman, who had graduated from Yale and become chair of Lehigh University's new civil engineering department in 1878. Other works commonly studied at the school in the early 1900s included the German city planner Reinhard Baumeister's *The Cleaning and Sewerage of Cities*, the U.S. architect William H. Birkmire's *Skeleton Construction in Buildings* and *Architectural Iron and Steel*, the French scientist and mathematician Jules Pillet's *Traité de stabilité des constructions*, and the U.S. engineer Frederick P. Spalding's *Roads and Pavements*. The works of one of Mexico's most famous nineteenth-century engineers, Francisco Díaz Covarrubias, were also widely circulated.[29]

Although most of their training focused on engineering, they did not completely ignore other subjects. They studied some elements of law and political theory. They were well versed in the creation and makeup of Mexico's 1857 constitution. Rolland and his compatriots would have been introduced to the French political economist Paul Beauregard's *Eléments d'économie politique*, which was a commonly used textbook at the time.[30] Beauregard examined economic problems in France in "a liberal spirit, not departing from the fundamental axioms of the orthodox economy," but he did "not accept absolutely the maxim of *laissez faire*."[31] His textbook focused on fairly orthodox and moderate liberal thought, with obvious influence from British proponents of free trade, but it also showed the influences of democratic socialists and progressives who hewed to more moderate principles between the doctrinaire free-market capitalists and socialists. Engineering students surely also came across more radical texts, such as those by Karl Marx and the Russian anarchist Peter Kropotkin, which were discussed with some frequency among professors and students in law, literature, and journalism.

As Rolland's generation of engineering students graduated and became more prominent in Mexico City intellectual circles, they interacted with people from different academic backgrounds and having more diverse perspectives, people who were increasingly antipositivist. This was especially true of law and humanities students who became critical of Mexican politics, survival-of-the-fittest ideologies, and to some extent the dominance of European philosophy. But Rolland and his engineering peers held more firmly to science-driven prescriptions for making improvements in society. They saw betterment as intrinsically intertwined with technological progress and improved material conditions. They often referred to themselves as apolitical, adopting that stance to appear scientific and to argue for specific policy changes while distancing themselves from the dangers of political confrontation.[32]

These young engineers and engineers-in-training nonetheless possessed a strong sense of duty. Working on projects throughout the city, they interacted with people who were often mystified but supportive of attempts to improve drainage, to increase access to water, to build homes and parks, and to reduce fire hazards. These men came from different levels of society but possessed immense ambition. Many of them took jobs in the government as mid-level planners and regulators, but a number of them aspired to become high-ranking advisors, like some of their mentors. These professionals bridged the worlds of everyday people and the most privileged members of society.[33] Rolland's generation of engineers genuinely wanted to improve conditions for all Mexicans, but in the process they acquired a hubristic notion that they were the only ones truly capable of bringing about an improved and modern society and that they thus deserved to be in positions of power.

Taking advantage of growing enrollments in Mexico City colleges, Rolland began teaching at the university level in 1905, accepting an instructorship at the Escuela Nacional de Agricultura y Veterinaria, or National Agriculture and Veterinary School. He taught courses on mathematics, drainage, and irrigation. He also lectured at the Colegio Militar, or Military College, and at his own university, the National School of Engineering.[34] He published his first writings: *Some Lessons on the Lifting of Polygons by Deflections* and *Lessons on Dams*. Unbeknown to Rolland, the brief texts would have a lasting influence. The first of these works would be republished in Yucatán in 1916, when former students under Rolland's direction were carrying out many of the agrarian reforms that became the hallmark of the Mexican Revolution.[35]

Virginia de la Garza Meléndez

It was during his time as a professor that Rolland began to seriously pursue romantic relationships. Following a failed attempt to marry a young woman named Luz Elvira Del Castillo, Rolland began a courtship in 1907 with a teenager by the name of Virginia de la Garza Meléndez. She was beautiful and extremely intelligent. She was seventeen, had a soft but proud face, and a sharp wit. Her eyes were rounder than Rolland's, and she had a small, round, youthful nose and a beauty mark to the right of her lips. She wore a gold necklace with an ornamental pendant. On the front side of it an inscription read, "The insignia of normal student Virginia de la Garza." On the back another engraving stated, "Medal of Honor given by Lic. Justo Sierra. Minister of Education. México, D.F., 1907."[36]

The necklace was a prestigious award. She had recently received it in a ceremony in the historic center's Arbeu Theater, the house of Mexico's Congress. President Díaz personally recognized her, members of the diplomatic corps applauded her, and Minister Sierra placed the emblem around her neck. She had graduated valedictorian from the Normal School for Teachers. In his speech Sierra praised her ability to transcend the "the inquietude of the concerns of the heart that fill the young and feminine mind with illusions." Directing his gaze toward de la Garza, he continued: "The Mexican woman has a great future that was not accessible in the past because they had been unfairly banned. But now you have every right and a complete show of support to enter the Temple of Science and Intellectual Culture, to study architecture, engineering, all career branches. . . . You are called upon to collaborate in grand works at the side of man and to be as God the Creator wanted from the beginning: the ideal companion to man."[37] Despite the patriarchal rhetoric, de la Garza represented an increased effort by Porfirian officials to educate young women, even if she was one of only a select few who truly possessed access to quality schooling. But that was not what was on her mind in the Arbeu Theater. This was her moment. She was proud. The crowd cheered her success and the ideal she stood for: progress.

De la Garza's education was the result of not only her hard work but also her place in society, something from which Rolland hoped to profit. She came from a prominent family. Her father, Miguel de la Garza Velazco, was a surgeon and gynecologist, one of the few in northern Mexico. His father before him had been a large landowner. Virginia's mother, Virginia Meléndez Rocio, descended from prominent women and Mexican officials. She and her husband had eight children; Miguel had also fathered at least three children out of wedlock.[38]

Rolland, too, had taken on the trappings of a successful urbanite. He still parted his hair to the left, cut short with a wave on the left side. But he put more effort into it. Rolland had always dressed as well as he could afford, but now he could afford better. During his courtship of de la Garza, he had invested in suits, ties, and shoes, which he kept well polished. His mustache had filled in. He began twisting the ends of it up the way other professionals did.[39] Rolland had found success building homes, assisting with the construction of waterworks in Mexico City, and developing methods of using concrete, a new trend in construction for which Rolland quickly became a leading proponent.

Rolland and de la Garza fell fast in love. They were attracted by their mutual intelligence and bravado. Her parents did not approve of the relationship. Stubbornly, the undaunted de la Garza married Rolland anyway. The couple became engaged in February 1908 and married shortly thereafter. In their engagement photo they both look youthful, especially de la Garza. Rolland wore a fine suit. De la Garza clothed herself in an all-white dress, her award pendant still proudly displayed on her chest.

The intense love they shared during their courtship quickly turned to conflict. They were both hot-headed. Rolland had an explosive temper, and he directed much of his attention toward professional advancement. Nevertheless they had their first child in mid-December 1908, a baby girl they named Enriqueta.[40]

Enriqueta's birth had been extremely difficult and, along with Rolland's constant working and fiery disposition, left Garza de Rolland miserable. The marriage changed Rolland. He had become increasingly repressive and a different person in public than he was in private. She complained that Rolland did not provide her enough money for household expenses. Late in her pregnancy before Enriqueta's birth, Virginia's mother came to stay with them. Modesto constantly argued with her as well, eventually kicking her out of his and Virginia's house. According to Virginia, this tension led to her having Enriqueta prematurely.[41] She thought he was a tyrant. He thought she was a brat. She filed for divorce.

The divorce never came to be. Rolland defended himself in a combative letter that shamed his wife into returning to their home. Modesto contended that Virginia had played her parents and him off each other, fueling divisions and using her family to prop up her attacks on him. Modesto went so far as to claim that her actions were killing her father, who was actually dying of stomach cancer: "When a woman like you has done all imaginable to get what she wanted, when a woman who, like you, has insulted me constantly, seeing an enemy in the one who gave everything for her, and when a woman has not hesitated to be false, saying lie after lie, even to the point of provoking the death of her own father, that woman does not deserve to be liked." Despite his cold-blooded tirade, Modesto resisted the divorce. He was hurt more than anything else. It was a matter of reputation and pride. In the end Virginia called off the divorce, pressured by Modesto and influenced by her dying father.[42]

In these formative years Rolland's personality became consumed by a need for control. He strove to control not only his spouse but everything around him. His vicious retaliation for Virginia's defiance may be a clue that he felt wounded by her and compelled to control her in order to tame his own hurt feelings. Rolland's fierce desire to direct the forces of nature, society, and politics through the science of engineering also seems, at least partially, to emerge from bruised pride and an attempt to transcend his heritage.

A Concrete World

Rolland was not alone in his determination to orchestrate nature and human organization; it was a common trait among Díaz's most capable advisors. Control and the desire for order were regular characteristics among engineers. As one later historian put it, "Engineers held strong convictions that through education and training they could control natural forces for human benefit, in effect combating the ills of the natural world through technology."[43] Engineers shaped the very nature of the earth, cities, and societies. They created networks of civilization. They were tentacle builders. Rolland's father had been a man trying to escape the trappings of civilization; Rolland became its tailored agent.[44]

Rolland had particular designs for his Mexico. He wanted it more concrete. Rolland's mentors, especially professors Marroquín y Rivera and Antonio Anza, were some of the first Mexican engineers to use concrete in large-scale projects, and they had a profound impact on Rolland's endeavors. Luis Salazar, the director of the National School of Engineering built a working relationship between the school and some of the new producers of Portland cement that had established themselves in the state of Hidalgo.[45] Concrete could be made cheaply and was relatively durable, especially when reinforced with steel rebar. Rolland, along with his friend Edmundo Cardineault, strived to improve upon methods of concrete construction, and Rolland would go on to earn some of the first patents on reinforced concrete in Mexico.[46]

Rolland and his mentors made extensive use of concrete. From 1906 to 1910 Rolland worked with engineers as they experimented with and incorporated reinforced Portland cement-based concrete into their projects. The material had been in use around Europe since the mid-1800s, but it had only come into vogue, even in the most industrialized nations, in the late 1890s and early 1900s.⁴⁷ It had been incorporated into the Gran Desagüe, or Grand Canal, the Díaz administration's massive drainage project designed to rid Mexico City of its destructive flooding. In a grand attempt to thwart nature—the capital was after all sitting on (and sinking slowly into) an ancient lake bed—the British firm Pearson and Son, along with prominent Mexican engineers, including Roberto Gayol y Soto, Miguel Ángel de Quevedo, Francisco de Garay, Blas Balcárcel, and Marroquín y Rivera, worked to rid the city of flood water and sewage while bringing in more water for drinking and fighting fires.⁴⁸ Workers cut through a mountain and built thirty miles of canals.⁴⁹ Similar to the Erie Canal in the United States, the Grand Canal and associated potable water projects became the training ground for a generation of engineers.

Pearson and Son, owned by the legendary engineering magnate Weetman Pearson, a man who would become Lord Cowdray and one of Britain's most successful engineers and business operators, had completed the Grand Canal and much of the drainage project by 1901, but Mexican engineers, including Rolland, continued to work with concrete and focus on bringing water into the capital. Starting in 1906, Rolland assisted Marroquín y Rivera on the Xochimilco–Mexico City aqueduct, Dolores water tanks, and the Condesa fire station. Rolland specifically helped build Pumping Plant No. 1 and the aqueduct from the Mexico City neighborhood of Chapultepec to the neighborhood of Condesa.⁵⁰ Some of the other sixty-three "engineers and employees" who aided in the potable water project included members of the Pani family—Alberto J. Pani, Arturo Pani, and Julio Pani—as well as Cardineault, Federico Cabrera, and Juan Francisco Urquidi.⁵¹ At the same time, Rolland built a number of other structures, including homes and an impressive seventy-two-foot-tall building near the famous Glorieta de Colón, or Columbus Roundabout. That building later became a restaurant called Shirley's.⁵²

In 1909, the same year he received his official engineering degree and fought the divorce case with his wife, Rolland began giving a series of lectures on reinforced concrete, which were well received by his peers. The following year Rolland published a brief text, *Cemento armado: Elementos de calculo* (Reinforced concrete: Elements of calculus). He had established himself as the leading expert on reinforced-concrete construction. *Cemento armado* would continue on through three editions, the last of which was published in 1948.[53]

The Apolitical Becomes Political

As if Rolland's life was not busy enough, he decided to immerse himself in Mexico's tempestuous political world. Many young engineers, like some of their mentors, viewed their work as important and apolitical, but they became critical of Díaz's advisors, cronyism, and the lack of genuine democratic practices.[54] Spurred by their strong sense of civic duty, engineers demanded attention in an increasingly technocratic government.

The years from 1906 to 1910 were tumultuous. Labor unrest grew, including at the Cananea copper mine in the northern border state of Sonora, where in 1906 a miners' strike for fair treatment and wages had been met with extreme violence, mostly by U.S. vigilantes from Arizona and by Mexican *rurales*.[55] The following year an economic crisis starting in the United States, often referred to as the Panic of 1907, dramatically affected Mexico. The Díaz administration became increasingly repressive in the wake of unrest. In 1908 Díaz had told the U.S. journalist James Creelman that Mexico was ready for true democracy and that he was not going to run for an eighth term. Parts of Mexico erupted with political fervor. And then Díaz, nearing eighty years old, changed his mind. He decided to run. It was like kicking a beehive.

Díaz faced not only the resentment of radical laborers and disgruntled peasants who had not benefited from engineering projects, railroads, and land speculation but also the disfavor of many of the very people his administration had provided training to—the proponents of modernity, people like Rolland. They organized into different groups that worked to increase their presence and their ability to engineer social change.

With a number of other engineers Rolland founded the Engineers' Club in 1908 "with a view of studying national problems." In Rolland's words "it was time to take part in public affairs." The club promoted itself as staunchly nationalist. It promoted the "Mexicanization" of the Ferrocarriles Nacionales de México, or National Railways, the development of increased infrastructure in the provinces, and "postal savings."[56] Rolland wanted more efficient communications systems, and he wanted them operated by Mexicans. For Rolland and his colleagues, Mexicanization meant the replacement of U.S. railway service personnel, especially in leadership positions, with Mexicans. The Engineers' Club contended that this shift would provide more positions for Mexicans and that it would better secure Mexican sovereignty.

The National Railways had a long, thoroughly twisted history with U.S. investors. Although the involvement of U.S. money in Mexican railroad schemes predated the Díaz era, it was during his dictatorship that American financiers had invested most heavily in building Mexico's railroads. By 1896 U.S. East Coast investors "owned 80 percent of Mexico's railroad stocks."[57] Most of the conductors, engineers, and managers were U.S. citizens. Due in large measure to criticism from a rising tide of Mexican nationalists and a real fear, even among Díaz's closest advisors, about U.S. expansionism, the Mexican government, via its treasurer, José Y. Limantour, moved to obtain a controlling interest in certain railroads in 1903. It bought a large number of shares in three railroads, which the Díaz administration and Speyer and Company, the government's U.S. business partner, merged into the National Railroad Company. In 1906 the government further consolidated its control, obtaining just over 50 percent interest in the National Railroad Company and a number of other lines, thus "nationalizing" the railroads and creating the National Railways, which was formalized in March 1908. Limantour also discussed totally Mexicanizing the rail industry.[58]

Yet the government still didn't control most day-to-day operations. Mexican state officials and U.S. financiers established two boards of directors, one in New York and the other in Mexico City. The Mexico City board worked as a collective intermediary between the Mexican government and U.S. capitalists, its role limited mainly to making suggestions. Americans appointed the directors and still controlled decision-making and the actual management of the railways.[59]

Edward N. Brown called most of the shots. He was a slender, mustached man fond of blunt statements. He had worked his way up the managerial ladder on railroads in the U.S. South. He got involved in Mexico in 1887, when he became assistant superintendent of construction for the National Railroad Company's line from Saltillo to San Luis Potosí. By 1902 he had become a vice president of the company. In April 1904, at the age of forty-one, he was elected president. He would hold the position until 1913.[60]

Rolland was not completely opposed to partnering with U.S. capitalists, but he argued that it was time for Mexicans to take up the mantle of leadership within Mexico. It is unclear if he knew exactly what he was getting himself into. The men who stood opposed to Mexicanization—the aforementioned U.S. investors and managers—were heavy hitters. They were the most powerful men in the railroad industry and some of the world's most powerful financiers.

Rolland also challenged the top leadership within Mexico. He signed on to a new political party, the *antireeleccionistas*, or Anti-Reelectionists, joining with a number of intellectuals, including lawyers, doctors, and engineers, to oppose organizations promoting Díaz's reelection. A number of the members of this first Central Anti-Reelectionist Club would go on to hold important government positions, and among those members were Rolland; Luis Cabrera, a lawyer; José Vasconcelos, an intellectual; Félix Palavicini, an engineer; and Francisco I. Madero, the soon-to-be revolutionary leader and president. At least seven of the other original members were young engineers. The professional children of the Porfiriato were raised on the history of liberal democracy in addition to notions of progress. They expected great change, technologically and politically, and they resented the old political order. According to the initial club signatories, its members came together because "reelection of public functionaries posed a serious danger to the democratic institutions ... and public spirit" of Mexico.[61] The Central Anti-Reelectionist Club announced that it would put up its own candidate for vice president and perhaps president as well, and it called for the creation of other affiliated Anti-Reelectionist clubs.

Joining this anti-Díaz group was a risky move. Díaz was an aging dictator, but he still wielded significant power and was known to take a dim view of direct opposition. And Rolland had been successful. He had much to lose. He attended high-society banquets, lectured at universities, and had married a woman from a respectable family. But Rolland wanted more. He placed his chips with the opposition, betting that it offered even better chances for advancement.

However, it was not only personal gain that motivated Rolland. He believed in the ideals he endorsed. Despite the Díaz administration's modernizing drive, the dictatorship had made a mockery of democracy, halting Mexico's ability to join the truly modern and successful nations of the world. Rolland would reflect five years later that he had joined the Anti-Reelectionists convinced of the "tremendous social inequality under the authority of capitalists and the clerical party" under Díaz.[62] Rolland had picked up a strong mix of liberal and socialist influences from his former mentors, peers, and a slightly younger generation of more radical thinkers in the capital. He was a cocky idealist. But as with many of the professionals who joined urban political movements against Díaz, Rolland did not promote violent revolution. That went against his order-oriented sensibilities. He naïvely hoped to "conquer a tyrant by persuasion."[63] The massive revolution that followed the 1910 election, the revolution Rolland helped initiate, was more than he bargained for.

2

The Reluctant Revolutionary

PORFIRIO DÍAZ, VICTORIOUS IN his seventh reelection, put on a lavish celebration that lasted an entire month. One hundred years earlier, on September 16, 1810, in a small church in Dolores, Guanajuato, Father Miguel Hidalgo had declared rebellion—the *grito*. It marked the beginning of Mexico's movement to gain independence from Spain. For President Díaz the festivities a century later honored more than independence; they reinforced his own regime. The president had actually kicked off the celebratory events on September 15, proclaimed as Porfirio Day, which celebrated his birthday. There was a grand parade. Representing a chronological narrative of Mexican progress, from feathered Aztecs to suited Porfirians, the participants marched metaphorically through time and into the central square, past the National Palace, President Díaz, and foreign dignitaries.[1]

During the rest of the thirty-day extravaganza the Díaz administration celebrated a number of grand public works projects. In addition to opening a mental asylum, hospital, and seismological institute, the government inaugurated the Xochimilco Potable Water Supply Works, the project on which Manuel Marroquín y Rivera, Modesto Rolland, and many others had worked so indefatigably. "Modern water" had arrived in Mexico City.

Even more than the new waterworks, the durability of the Porfirian administration was a façade. The centennial celebration in Mexico City just barely hid a problematic political situation—a president past his prime and growing displeasure in the vast regions outside the capital. There had been a serious drought in parts of the countryside, labor unrest, growing antiforeigner sentiment, and the usurpation of communal lands around villages by surveyors and hacienda owners. In the capital many middle-class professionals, including Rolland, had opposed Díaz's continued rule. The election had been contested and corrupted by fraud.[2]

It is unclear what Rolland was doing at this time. The tense political atmosphere had forced most Anti-Reelectionists into subdued silence.[3] Rolland never discussed the immediate aftermath of the 1910 election in his writings despite being one of the original members of the Anti-Reelectionists. Instead of attending the extravaganza that inaugurated the waterworks, he built houses and patched up his relationship with his wife. Garza de Rolland was pregnant during much of 1910 and 1911. Their second child, Martha, was born in January 1911. Their third child, Alberto, was born a year later, in February.[4]

Things would not remain quiet for long. Allowed his freedom following the election, Francisco I. Madero fled to Texas, where he declared rebellion against the Díaz administration. The Mexican Revolution had begun. Madero's supporters fared poorly at first, but thousands of people, especially in rural communities, soon rallied to his banner, more often than not out of a desire to effect change in their local circumstances than any particular love for Madero. To the surprise of many Mexicans and foreigners alike, Madero's forces toppled the dictator within a year.

After Díaz went into exile, Rolland once again engaged more vocally in public affairs. He broadened the endeavors he had begun during the years before, experimenting and building with reinforced concrete, constructing homes, filing patents, and providing public and closed lectures. He continued to organize engineers in the capital to lobby for scientific planning and the Mexicanization of railroads. He became savvy in using Mexican newspapers to influence policy. Rolland also turned some of his attention back to his home region of Baja California, teaming up with other Bajacalifornianos residing in the capital to pressure the new government to provide residents of the peninsula a greater voice in governance, education, and development even while connecting Baja California more firmly to mainland Mexico through new infrastructure. Much to his frustration, continued turmoil and political resistance stymied many of his endeavors.

Madero's Rebellion

Before the rise of the Anti-Reelectionists, many Mexicans had looked to Gen. Bernardo Reyes, the governor of Nuevo León and former minister of war, as the successor to Díaz. Reyes had especially strong support in northern Mexico. He possessed a broad network of promoters and looked to be a serious contender, though he never declared his candidacy himself. He was a loyal Porfirian who had been successful working with local capitalists to industrialize the city of Monterrey.[5] Along with the treasurer, José Y. Limantour, Reyes was one of the most powerful men of the era, but the two cared little for each other. Reyes quarreled with many of Díaz's *científico* advisors, which turned many of their enemies into his adherents. Díaz respected Reyes, but he also considered him enough of a threat to send him on ambassadorial duties to Prussia in November 1909, an assignment that amounted to exile during the election.[6]

Madero, the presidential contender who capitalized most on Reyes's departure, was the son of a wealthy and influential family. He was short and well groomed. He had attended universities in France and California before settling down to run one of his family's large farming operations in the northern state of Coahuila. Reform-minded intellectuals had influenced him greatly, and he worked sincerely to improve the lives of many of the workers under his management.

Madero initially became involved in politics in the early 1900s, first locally, then regionally, and then nationally. He rose to political prominence following the spread of his book *The Presidential Succession of 1910*, which had been published in late 1908. He and his liberal policy prescriptions—no reelection, effective democracy, and freer governance at the state and municipal levels—won traction following the departure of Reyes. Madero gained the backing of many former Reyes supporters in northern Mexico and petit bourgeois intellectuals, including Rolland, in towns and large urban areas.[7] As the historian Alan Knight writes, it was during the 1909–10 political campaign that a politicized "middle class made their decisive contribution to the gestation of the Revolution."[8] Forming an important bloc of the party that surrounded Madero, Rolland and his associates were critical in igniting political change. They would go on to play a significant role in the outcome of the revolution and the shape of twentieth-century Mexico.

Once Madero had obtained the Anti-Reelectionist nomination for the presidency, he undertook a nationwide tour. It was the first time in the country's history that a presidential candidate stumped so broadly. The campaign gained momentum as it progressed, but it faced increased intimidation from Porfirian officials. Madero was arrested on bogus charges of sedition on June 13 after a disorderly rally in Monterrey. Initially hauled to a jail in San Luis Potosí, Madero sat out the remainder of the election restricted to the confines of the city. Just weeks following Madero's detention, Díaz proclaimed electoral success by an unreasonably wide margin.[9]

It did not take long for Madero to complete his transition from reformer to revolutionary. After being released from detention with help from his family and Porfirian friends, he fled to the United States in early October. Disguised as a laborer, he crossed the border at Laredo on October 7, 1911, and formed a temporary base of operations in San Antonio, Texas. There, he drafted the Plan of San Luis Potosí, denouncing the actions of Díaz, declaring the presidential elections a fraud and hence null and void, and calling for armed rebellion to begin on November 20 at 6:00 p.m. Madero declared himself provisional president.[10]

The uprising started poorly. Porfirian agents and soldiers uncovered plots and arrested Madero sympathizers. Predictably the movement floundered in the hands of urban intellectuals, who lacked both means and the inclination for armed struggle. Middle-class professionals had helped solidify Madero's candidacy, but almost none of them did anything to seriously prepare for a violent revolution, which was never their intention. Madero found little support upon crossing the border the day before the uprising was set to begin, which sent him back to San Antonio for a brief period. The engineers, lawyers, and doctors proved to be poor revolutionaries. They were timid. Possessing no taste for physical combat, they either did nothing or they waited for "official" orders.[11]

Despite its embarrassing start, Madero's revolt ultimately prevailed. This success had little to do with the Anti-Reelectionists in the cities; it was men and women from the rural countryside, largely in the northern states of Chihuahua and Durango, who won Madero's victory. In addition to these northern warriors, other largely autonomous insurrectionaries fought for lands in the small central state of Morelos. These fighters were not "civilized" intellectuals.[12] They rebelled because of grievances about access to farm and ranch lands, mistreatment, and personal grudges. Localized uprisings sprang up across much of Mexico, nominally taking on the name of Madero, but Madero had little control over them. Joining with some of the more successful rural military leaders of the nascent rebellion in the north—Pascual Orozco, José de la Luz Blanco, and Doroteo Arango (Francisco "Pancho" Villa)—Madero eventually surrounded the border city of Ciudad Juárez in mid-April 1911. He and his family members worked with Porfirian leaders to come to terms. After it became apparent that Madero planned to pull back on May 8 after receiving limited concessions, his army attacked the city against his orders. To the surprise of outsiders, the rebels took the city. Insurrectionists in Morelos under the direction of community leader Emiliano Zapata took that state's capital, Cuernavaca, soon thereafter. Facing increased pressure, uprisings across the country, and a serious toothache, Díaz resigned. He set off for France on a steamship in early June.[13] Madero had won, but the once political rebellion had turned into a social and largely rural revolution that the fledgling president-to-be struggled to contain.

Rolland's Return

Rolland popped his head up from the sea of revolution after Madero's triumph. During what appeared to be a surprisingly short revolution, Rolland owned Reinforced Concrete Works, drawing little attention to himself while he built homes and reinforced-concrete products. Following Díaz's ouster, the new undersecretary of communications and public works, Manuel Urquidi, approached Rolland about participating in a government commission tasked with studying and promoting increased production using reinforced concrete. Rolland agreed, volunteering his services for free. His private workshop was doing well. In addition to his business constructing residential homes, he had designed a new type of reinforced-concrete water tank. It possessed a meshed-metal skeleton and was, according to Rolland, more efficient and cheaper to make than other concrete water tanks. While Madero experimented with deals and appointments, Rolland was building and experimenting far more successfully with reinforced concrete.[14]

But Rolland's private business never fully satisfied him. He helped form two organizations in mid-August 1911: the Anti-Reelecionista Club Francisco Díaz Covarrubias, which consisted entirely of engineers, and the Club Progresista Californiano. For Rolland, both groups were intertwined. Each one worked to strengthen Mexican sovereignty and nationalism. He spoke often that Mexico needed more east-west development to counter the historical expansion of U.S. companies building north-south railroads, which had siphoned resources from Mexico. Further, Rolland stressed, Mexican engineers should be in charge of these infrastructural projects. His passion and drive came at least in part from his desire to improve his home territory of Baja California. Rolland contended that underdeveloped regions were in need of crucial attention and were especially in need of infrastructure connecting them to the heart of Mexico. At the same time, local municipalities needed more freedom in electing their own officials without the central government forcing leaders from Mexico City on them. Mexican development had been too uneven, leaving behind certain regions. Instead of placing officials from the national center in regional posts, the government would be better off increasing education and developing specialists in the regions themselves, allowing people to improve their own lives, and then connecting them through infrastructure and trade, not political domination.

The Club Francisco Díaz Covarrubias elected Rolland as its president and was more or less an extension of the Engineers' Club but with a new mantle consisting of Madero's slogans: "effective suffrage, no reelection, antipersonalist politics, diffusion of civil governance, abstraction of religious ideas, and the spread of democratic ideas most adaptable to the country."[15] The members represented a nationalist wing of Mexican engineers, individuals who continued to push for replacing foreign specialists and contractors with Mexicans.[16]

Most of their work was not focused on anticlericalism or the democratization of the Mexican political system; instead it homed in on infrastructural development and economic nationalism. Even though Madero was not yet president—a former Porfirian ambassador and cabinet member, Francisco León de la Barra y Quijano, was acting as interim president while Madero waited for "official" elections on October 15, 1911—Rolland helped reignite the campaign to further "Mexicanize" the National Railways. Rolland and his allies worked particularly hard to replace U.S. members of the company's leadership, especially the president and vice president, with Mexicans.[17]

Carlos Meza, a prominent Bajacaliforniano, headed the Club Progresista Californiano. Rolland was the secretary. Initially ten members strong, the group attempted to organize Bajacalifornianos in Mexico City in order to promote the peninsula. They appealed to Interim President de la Barra to establish commissions to study ways to better exploit the peninsula's natural resources and improve the region's irrigation and communications systems. Their specific recommendations were to separate the civil and military authorities and make the new civil authority a Bajacaliforniano, to increase Mexican colonization of the peninsula, to build new communications links, to revise contracts with foreigners, to better develop and protect marine resources, and to reorganize the territory's education system.[18] Although de la Barra, and later Madero, were willing to listen to the Club Progresista Californiano, constant turmoil hindered any real progress on these issues.

Despite the limited success of the organization, the Club Progresista Californiano's actions show the group's attempt to increase nation-state construction and regional development. Rolland's nationalism pushed him to unite Mexico more firmly through infrastructure, colonization, and education, but he and the rest of the Club Progresista Californiano held on to strong notions of regional independence, a remnant of nineteenth-century federalism that was likely influenced further by progressives who were pushing for greater municipal reform against corrupt political practices in the United States. Rolland wanted to shape policy at the national level, but he wanted those policies to include a Baja California peninsula with democratic communities.

Mexicanization

In late August 1911 Rolland rebooted his Mexicanization campaign through the Club Francisco Díaz Covarrubias. The organization was not alone in its crusade. Unionized railway workers were threatening to strike, providing the real energy behind the movement.[19] They demanded better treatment and better jobs. Rolland jumped into the fight by first creating the club itself and then by crafting a plan to gain widespread public support in order to influence lawmakers and the boards of directors for the National Railways. The boards held their annual meetings in early October. It's unclear if Rolland's actions arose from a sincere concern for the workers or if he saw the unrest as creating an opportunity to try to place specialists like himself into positions of greater power. His rhetoric focused on fighting for the people, but his actions centered on keeping order and making changes in positions of leadership. Perhaps he believed he was genuinely helping both workers and engineers like himself, but he definitely pushed for more gradual changes than radical labor activists did.

Rolland and the other club members found their greatest success in their ability to influence public opinion through the press. They undertook a publicity campaign in the capital, publishing articles in various Mexico City newspapers, including *Nueva Era*, *Diario de Hogar*, *Elektron*, *El Imparcial*, and *El País*. Instead of attacking domestic political enemies, Rolland and his associates reached out to them. The leaders of the Partido Católico, Centro Reyista (Reyes had returned from exile and decided to run for the presidency), Partido Nacionalista Democrático, the Partido Independente de Jalisco, and the Partido Constitucional Progresista (Madero's new party) all joined Rolland's Mexicanization plan. He obtained the support of many of the capital's engineers and a number of other influential intellectuals. Considering that Rolland was not a fan of the Catholic hierarchy, his willingness to work with the Catholic party suggests that he was not opposed to putting practical considerations ahead of ideological beliefs. Rolland also wrote essays addressed specifically to National Railways workers and published their responses in newspaper articles. He obtained more than four hundred signatures to give to members of the Congress and the National Railways boards.[20]

Rolland presented a clear goal of increasing the presence of Mexicans in the National Railways directorship, with a particular focus on replacing the company's U.S. vice president and president with Mexicans. The New York board consisted of nine New Yorkers. The Mexico City board consisted of twelve men from Mexico City, but according to Rolland, only seven of them were actually Mexicans. Almost all of the National Railways department chiefs and supervisors were Americans, and so too were most conductors and engineers. To his taste, these facts demonstrated a lack of concern for national interests in the management of the so-called National Railways.[21]

Rolland's general thoughts on foreign investment and workers' rights were more opaque. He specifically stated that the Club Francisco Díaz Covarrubias was not carrying out an "antiforeigner campaign," but then Rolland attacked U.S. management of the railways and American employers as abusive fearmongers. Rolland promised to "expose concrete examples of abuse" by Americans, and he vowed to help fulfill revolutionary democracy by aiding people at the lower end of the social ladder. He promised railroad workers that he acted on their behalf, yet he did not want to cause massive "shocks" by attempting to replace every "average Joe" U.S. worker. When railway workers planned strikes in November following Madero's election, Rolland reprimanded them. He told people that he was not inherently against strikes, but he argued that the timing was not right.[22] The workers were not strong enough and lacked organization. The Madero administration was too new and too weak to support strikes against powerful foreign interests like the U.S. railway financiers. Rolland recommended patience and suggested union members focus on strengthening their skills. He thought that radical protest would prove to be disorganized, dangerous, and counterproductive. It was just as likely that he really opposed the idea of striking workers but carefully couched his opposition in terms of tactical concerns so as not to appear reactionary.

His campaign was less dangerous, perhaps, but it failed to place Mexicans into the presidency and vice presidency of the National Railways. Rolland's initiative, along with similar pressure coming from other Mexicanization groups, did, however, garner significant attention. The leadership of the National Railways publicly proclaimed that it would "deliberate conscientiously on this very delicate problem in order to find a solution that would satisfy public opinion."[23] However, the changes that came about in October were superficial. A few Madero supporters, including one of the president's brothers, Gustavo Madero, took over positions from their Porfirista counterparts.[24] But as Edward N. Brown, still president of the National Railways, smugly put it, "No particular significance attaches to any of them [the changes in the directorate]. The policy heretofore of the system's management as regards to expansion of the properties and new construction work, as well as new financing[,] will continue to be carried out as before." According to Brown, the Mexicanization campaign had been "exaggerated . . . as to its activeness and bitterness with which the movement" had been carried out. The company would continue with its policy of "maintaining present incumbents, regardless of nationality, on the basis of general efficiency and demonstrated qualifications, with promotions to be made on the same basis, when vacancies occurred."[25] The policy may have sounded fair, but the U.S. financers largely saw Mexicans as incompetent and less desirable, which meant that Americans would continue to hold most of the leadership positions.

Instead of wanting to pick fights, the Madero administration was far more interested in calming foreign investors. It was not willing to challenge outright the railroad giants of the United States. Brown "spoke highly of Madero," stating that Madero "would welcome the investment of foreign capital looking to the development of the country's resources."[26] Rolland and his colleagues met stiff resistance from another of Francisco Madero's brothers, Ernesto Madero, the new secretary of finance, and his undersecretary, Jaime Gurza. Ernesto made no serious attempt to push for significant managerial changes and gave Brown "all possible aid and encouragement."[27] Reflecting back later, Rolland squarely blamed the failure of his Mexicanization campaign on Madero and Gurza: "Nearly all of our efforts were shattered by the Secretary of Treasury, headed by Messrs. Ernesto Madero and Jaime Gurza."[28] Rolland had underestimated the power of the U.S. railway financiers and the reluctance of the Madero administration.

Yet the Mexicanization campaign was more successful than Rolland later characterized it. Mexicans did not obtain the top positions, but a transition from U.S. management to Mexican management in the lower and middle sectors of the company took place in 1912. This move was at least partially influenced by Rolland and the Club Francisco Díaz Covarrubias, which continued to publish letters listing well-qualified candidates whom the members deemed fit for supervisory roles.[29] That Rolland later emphasized the failure of Mexicanization while failing to acknowledge the victory for Mexican workers striving to obtain lower management positions reinforces the image that he was ultimately more about those who organized the railways at the top than about those who actually worked them.

U.S. railway workers in Mexico did not take the news well. Alleging favoritism toward Mexican workers from the Mexican government, U.S. conductors and engineers responded by threatening to strike in April 1912. They wrote inflammatory letters to the U.S. embassy, where Ambassador Henry Lane Wilson, who avidly despised Madero and seemingly most other Mexicans, relayed exaggerated claims of Mexican abuse of American workers to Philander C. Knox, the U.S. secretary of state. According to Wilson, the Mexicanization campaign had purposefully unleashed a wave of anti-Americanism that would doom the National Railways. He had little faith in the ability of Mexican engineers and laborers: "In the event that the Mexicanization of the National Railways is persistently and thoroughly carried out by the Mexican Government the deterioration of these great properties must almost inevitably follow."[30] To Wilson, Díaz had been a capable leader, but in his view the revolutionaries possessed little skill and were led by an incompetent president.

From mid-April to May hundreds of U.S. National Railways employees went on strike.[31] The Madero administration, pressured by workers and groups, including the Club Francisco Díaz Covarrubias, rejected a series of demands made by U.S. engineers and conductors. Among their demands was that the ratio of American to Mexican workers be held at the April 1911 level, that Americans obtain three-year contracts, and that the government reverse a recent order that the operations of the National Railways be carried out in Spanish. Ambassador Wilson had found the Spanish-language requirement especially unreasonable and "obnoxious."[32]

The Mexican government responded firmly. It rejected the ultimatums, finding them "intolerable."[33] It fired a large number of the strikers after they failed to return to work in a "reasonable time," and it then expelled them from the country. Many of the American engineers and conductors, "nearing the age limit of service[,] . . . could not hope to obtain work in the United States."[34] U.S. president William H. Taft reached out to the heads of more than eighty U.S. railroad companies, asking them to hire returning engineers and conductors, or at least the ones who were lingering in El Paso.[35] The National Railways continued to operate at a profit for the remainder of the year, despite the reduced number of Americans and despite the prevailing belief of Americans that the system would fall into ruin under Mexican management. However, continued rebellions—aftershocks of Madero's call to arms—threatened to throw the railroads and most of Mexico into disarray.

Continued Unrest and Madero's Fall

While Rolland and others were clamoring for attention to specific regions and for the Mexicanization of the National Railways, a number of rebellions menaced the very existence of the new Madero administration. Madero fought for political change, but he was hesitant to enact rapid social transformations. Instigated by exiled comrades and foreign radicals from Southern California in the United States, rebels associated with the anarchist intellectual Ricardo Flores Magón had risen up in northern Baja California and in the United States and pushed southward. The Madero government put down the last remnants of this revolt in the summer of 1911. Many farmers fighting for lands and urban laborers demanding better wages and opportunities found the Madero administration too much like the government it had replaced. Madero had rebuked revolutionary leader Emiliano Zapata's insistence that portions of hacienda lands in his home region of Morelos be turned over to villagers immediately. An infuriated Zapata took up arms again. He and his advisors drafted the Plan of Ayala on November 25, 1911, which called for the overthrow of Madero and put Pascual Orozco in place as the new revolutionary head. Orozco had not asked for the job, but that did not stop the Zapatistas from giving it to him.[36]

Orozco, Madero's former military chieftain, was not quick to respond, but he too was unhappy with the new order. Upset with Madero about not obtaining a high political position in exchange for his military service, Orozco finally decided on rebellion, which began on March 3, 1912. He had an army at least six thousand strong and the support of a motley consortium of wealthy landowners, radicals, and northern rancheros. After Orozco delivered serious defeats to Madero's forces, the new president turned to the old Porfirian army, specifically Gen. Victoriano Huerta. Known for his ruthlessness and military skill—he had already provoked bitter hatred among the Zapatistas—Huerta helped turn the tide against Orozco. Madero also relied on state militias to battle the insurrectionists, including a small militia led by Álvaro Obregón Salido in the northwestern state of Sonora. Obregón's success placed him on a path that would turn him into the revolution's most capable general and eventually the country's president.[37]

Reactionary movements also broke out, the most important being the rebellions of Bernardo Reyes and Félix Díaz. Reyes, who had exiled himself to the United States following his failed effort to gain the presidency, attempted a coup after crossing back into Mexico in December 1911. The Madero government had gotten wind of the plot beforehand. Contrary to his expectations, Reyes's movement failed to gain much support from the army or the public. Deflated, he turned himself in on Christmas Day, and Madero officials brought him to the Santiago Tlatelolco prison in Mexico City. Díaz, a former member of the Congress, a military officer, police chief, and nephew of Porfirio Díaz, rebelled the following year in the port city of Veracruz. He too thought the military would rise to his aid. He too was disappointed and quickly captured. After first holding Díaz in Veracruz, the Madero administration moved him to a different prison in Mexico City. The transfer pleased anti-Madero conspirators in the capital, who began plotting a new coup, one that would involve freeing Díaz and Reyes. Facing near constant uprisings, Madero and his close associates became overwhelmed with putting out the militant flames they had ignited, leaving little time to consolidate their rule and to complete projects.[38]

Madero, for example, told the Club Progresista California he agreed that Baja California desperately needed social reforms and development but that he would not be able to devote serious attention to these issues until other, more urgent problems of his nascent government had been resolved. Much to Rolland's disappointment, it became apparent that Madero had little control over the peninsula or anywhere else, and, despite some feeble attempts, nothing much was accomplished in the territory.[39]

Meanwhile in the late spring of 1912 the striking American railway workers sent letters to the U.S. embassy, stating that they would be more than happy to operate trains for the U.S. military if the American government decided to invade Mexico. With the outbreak of rebellions and what many U.S. officials deemed "anti-American" sentiment, U.S. government officials did weigh the possibility of intervention. U.S. business leaders had invested heavily in Mexico during the Porfirian era; they had a lot to lose. Most top U.S. government officials, however, were leery of the idea. But the threat was enough to convince many Mexican officials and intellectuals to prepare Mexico's defenses.[40]

Rolland's actions reinforced his centrist position within the Madero camp. He was passionately nationalist and anxious to have himself and other professionals who had been trained during the late Porfirian era to serve in positions of power, but he fundamentally disapproved of mobilizations from below and violations of the law. For example, Rolland joined a number of other engineers, professors, doctors, and military leaders led by Gen. Gerónimo Treviño to form the Agrupación Democrática Pacificadora Nacional, or National Democratic

Peacemaker Group. Their goals were to stem the growing violence in Mexico, to prevent its "sons from taking up arms" against each other, and to better unify Mexico against external threats. According to this group, "The patriotic action of the men elevated to power by the Revolution had been succeeded by our fiery Latino temperament, encouraging many of our brothers onto the battlefield, to fratricidal struggle, with a view of reaching by violent means what can and must be reached by order and harmony under the law." The organization promoted national solidarity beyond "distinction of political ideas, without distinction of religious credos, and without distinction of classes and categories."[41] It appears they would have been happy to have promoted the old Porfirian motto of "order and progress" if it had not been tainted by its connection to the former dictator.

But Rolland was not a pacifist. He, along with many of his National Democratic Peacemaker associates, formed a "provisional directory" to help Mexico avoid disgrace in case the U.S. invasion that American railway officers so ardently called for actually occurred. Rolland had been teaching at the Military College more regularly, which gave him some familiarity with the army. Claiming once again to be without "political coloring," the directory called for the formation of other patriotic groups based on particular skill sets so that they could receive "indispensable military instruction" on how to improve the mobilization of military forces and ambulances, speed up communications, and strengthen fortifications. Rolland and other involved engineers focused on how to obtain ammunition, arms, and manufacturing equipment and how to expand nationalist propaganda. Willing "compatriots" met with the engineers at the School of Arts, among other locations, to be organized into separate groups and receive training.[42]

The U.S. invasion never came; instead Ambassador Wilson helped overthrow Madero from within Mexico. On February 9, 1913, plotters led by two generals, Manuel Mondragón and Gregorio Ruiz, freed Reyes and Díaz from their respective prisons and then marched on the National Palace. Despite initial success in taking the National Palace, forces loyal to Madero led by Gen. Lauro Villar took back the building. Villar's troops shot and killed Reyes and many of his supporters as they attempted to retake the building. Díaz regrouped his forces at the Ciudadela, an old fort and armory in the capital. Villar, injured in the fighting, could no longer continue in charge of Madero's forces. The president fatefully, and against the wishes of some of his closest advisors, chose General Huerta to lead the movement to quash Díaz. In return Huerta turned on Madero. After a battle now known as the Decena Trágica, or Ten Tragic Days, Huerta made a deal with Díaz and captured the president and his vice president, José María Pino Suárez, and forced their resignations. Wilson helped broker the deal between Díaz and Huerta, consolidating the coup. Huerta took the presidency, while Díaz helped select the cabinet and was promised a prominent role in future governance. Soldiers, likely under Huerta's orders, executed Madero and Pino Suárez on February 22 outside the Lecumberri prison.[43]

Rolland had been active in combating anti-Madero forces during the Ten Tragic Days. Alongside Alberto Pani, Manuel Urquidi, Juan F. Urquidi, Efraín R. Gómez, Luis Salazar, and other engineers and technicians, Rolland helped provision troops, establish telephone communications, and clear the streets of debris and dead bodies. The overthrow of the president was a crushing blow to Rolland and most of his other former classmates who had supported the Madero revolution.[44]

In retrospect Madero's defects appear obvious. He was too slow to make the serious changes demanded by so many of those who had risen up under his banner, and he was too willing to compromise with his opponents. In the words of Carlo di Fornaro, a well-known Italian journalist and cartoonist living in New York who wrote extensively on Mexico and with whom Rolland would soon work, "When Madero came into power he tried to conciliate all his enemies, before he had really achieved the principles of the revolution. He attempted to compromise with the porfiristas, the cientificos, the reyistas, the felicistas, the vazquiztas, the zapatistas, the clericals, the militarists, the landowners, the foreign interests; he offered the olive branch to all, and he was repaid with the assassin's bullet. He had mistaken family loyalty for political loyalty and was therefore accused of nepotism and he was abused as incompetent and weak."[45] Good intentions were not enough to secure Madero's government. His loyalty to law, order, and methodical progress cost him support from those more eager for real change. His reliance on the old Porfirian military cost him his life.

Following the Ten Tragic Days, Rolland returned to the Military College. As he walked through the rubble-filled streets, his mind burned. He knew this would be the last time he talked to his students. Entering the classroom, he set down his hat and jacket. He told the students that Huerta would use the army, young cadets including themselves, to kill their own people, "that they would be the instrument of a traitor to shed the blood of Mexicans."[46] That afternoon Rolland was fired.

3

A Mexican Progressive

NOT LONG AFTER THE coup, some of Victoriano Huerta's soldiers pulled Modesto Rolland out of his business office. They had been watching him. They brought him to one of Mexico City's modern marvels—the penitentiary—and he was placed in solitary confinement, remaining in this "dark dungeon" for a month.[1]

The experience shook Rolland, altering the course of his life. Lucky for him, Rolland had powerful friends. These colleagues talked Huerta's secretary of *gobernación* (roughly equivalent to a minister of internal affairs), Manuel Garza Aldape, into meeting with Rolland. Rolland convinced the minister that he was not a threat, or at least not so dangerous that he could not be released. Back on the capital's streets, Rolland remained paranoid, and with good reason. People he knew had not been so lucky. They remained in jail or had been re-arrested. For Rolland it became harder to find work. His once prosperous "business affairs were shattered; every move was constantly watched."[2]

Like many other associates of Madero, Rolland fled Mexico, leaving his wife and four children behind. Virginia Garza de Rolland had recently given birth to another son, Jorge. Fleeing to the United States was in Rolland's words "the history of thousands of men in Mexico. Thousands of families remained until they had nothing left to live on, and even the women were in danger of being put in jail, as many were." With some difficulty, Rolland stowed away on a ship in the port of Veracruz "as contraband." He pretended to be a table steward before disembarking at some U.S. port.[3]

Civil war broke out shortly after Huerta's usurpation of power. The forces opposed to Huerta that loosely coalesced around Venustiano Carranza, the governor of Coahuila, called themselves Constitutionalists. They promised to overthrow Huerta and to resurrect the revolution that Francisco Madero had started. The fighting would drag on for seven years as the Constitutionalists themselves fractured in the face of their own success. In October 1914 the forces of the popular revolutionary leaders Pancho Villa and Emiliano Zapata broke away from the Constitutionalists, forming the Conventionalists. The Conventionalists and the remaining Constitutionalists fought each other, pushing Mexico into the most destructive years of the revolution. Ultimately, after much bloodshed, two invasions by the United States, complications stemming from World War I, and much persistence, the Constitutionalists (for the most part) had prevailed by the end of the 1910s. The revolution, however, would not be more firmly consolidated until Gen. Álvaro Obregón, Carranza's ablest general, overthrew him in the run-up to the 1920 election, the election that was supposed to be the first peaceful presidential transition since Madero's assassination.

The exiles who fled Huerta's Mexico played an important role in this revolutionary drama. They were mostly urban professionals and intellectuals, not soldiers, but they became important agents of the various revolutionary factions. Many of them remained in their host country for substantial periods of time, the majority of them residing in the United States, though others went to Canada, Europe, and other parts of Latin America.

Rolland became a champion of the Constitutionalist cause, and he would spend much of his time between 1913 and 1919 going back and forth between Mexico and the United States. His time in the United States proved critical to the development of his own ideas and important to the success of Carranza's forces. Rolland and a number of colleagues created intellectual bridges between influential American progressives and the Constitutionalists. Rolland in particular would become a capable propagandist and diplomat, swaying U.S. public opinion. In turn U.S.

progressive ideals had a profound impact on Rolland, influencing his policy prescriptions and engineering projects. Rolland formed most of his beliefs about educational and municipal governance at this time. He crafted agrarian reform policies, briefly obtained a position as a high-ranking communications official, and became involved in a Constitutionalist faction that remained loyal to Carranza but challenged Carranza's conservatism. The bonds formed in this clique would reverberate in the Mexican government throughout the remainder of Rolland's long career.

Designing Mexico from the United States

While still in prison in Mexico City, Rolland had told Minister Garza Aldape that he was not a conspirator against the Huerta government; he lied. Another explanation might be that if Rolland did not seek Huerta's downfall before being jailed, which his speech to military cadets suggests, he did so fervently after the incarceration. Rolland was not alone; dozens of Mexican professionals fled. But they had no desire to remain abroad. Their goal was to topple Huerta.[4]

In Mexico, too, plenty of people remained interested in fighting Huerta. Although Huerta had found legislative loopholes to provide a semblance of legality to his ascension, his presidency clearly made a mockery of Mexico's constitution and the spirit of the law.[5] Many members of the Congress and governors reluctantly fell in line with the new order, but the legislature and governor of the northern state of Coahuila called for resistance. The state's governor, Carranza, was a landowner and a former Porfirian senator, but he was also a stalwart of the Madero Right and a reformer. After some initial hesitation he called on other governors to rise against Huerta. At first he found few willing to join his cause. They feared losing. José María Maytorena, the governor of Sonora, commanded a large number of loyal forces. But fearing the destruction of his state and personal wealth, he fled across the U.S. border into Arizona to plot from there. Sonora's legislature, however, joined in condemning Huerta.

Carranza continued his revolt, issuing the Plan of Guadalupe on March 26, 1913, which called for the overthrow of Huerta, respect for constitutional law, and free elections. The lack of competing leadership made him the de facto leader or, as he titled himself, the "First Chief" of the Constitutionalist forces.[6]

Carranza's plan spoke a language Rolland understood, and he quickly resolved to meet with the First Chief. Along with other exiles, he made his way to the U.S.-Mexican border through El Paso and back into Mexico at Juárez. There he had his first encounter with Carranza. The First Chief left a marked impression on callers. He was tall. His goatee was long and white, and his mustache extended beyond the boundaries of his face. His dress was formal but simple, and he wore small, round, thin-framed glasses. In his mid-fifties, he was more than two decades older than Rolland and nearly three decades older than most of the would-be revolutionaries who took positions under him. Carranza was intelligent, ardently nationalist, and erudite if quiet, and he had a patrician air about him.

Carranza embraced educators and engineers eager to join his cause. Engineers knew about making munitions, running trains, and constructing public works. They were systems builders, and they would be essential to conducting war and reconstructing Mexico. After a long discussion on agrarian issues, liberal politics, education, and municipal governance, Carranza asked Rolland to return to the United States. The First Chief wanted to take advantage of Rolland's English-language skills, passion for reconstruction, and experience as an instructor.[7] Carranza tasked Rolland with studying U.S. schools and municipalities. He wanted his future government—he already acted confident of his victory—to possess clear plans on how to address important issues of education and city governance. Rolland welcomed the chance to return to New York City and to become an insider in Carranza's movement.[8]

Rolland thus "joined a body of students of administrative service" that Carranza was forming in the United States and Europe.[9] Carranza understood that the war for Mexico would be an international affair. He needed support and supplies from abroad. His forces required eyes, ears, and voices to report on foreign public opinion and to shape that opinion. He needed new ideas and models for infrastructure and reform. Under the direction of Carranza's secretary of internal affairs, Rafael Zubarán Capmany, Rolland joined fellow engineer Francisco Urquidi, who had been the consul in New York during the Madero administration and now operated as Carranza's "commercial agent" in the city. Other prominent Carranza advisors, including the already renowned lawyers Luis Cabrera and Juan Neftalí Amador and engineers Manuel Urquidi and Juan Francisco Urquidi (Francisco's brothers), took posts in places like Washington DC, New Orleans, San Francisco, and El Paso.[10]

Rolland began his investigation into U.S. schools in the spring of 1914, continuing his study well into autumn. He worked with the U.S. Department of Education and local education officials and set about studying schools in Wisconsin, Massachusetts, and New York. Carranza and Rolland's ultimate goal was a free and compulsory school system built on the most progressive models in the United States.[11]

The schools in the United States impressed Rolland. Reflecting on his experience, he wrote that "the soul of this nation palpitates in its schools. There the body and mind are fortified. These things are facts, not theories, in the American schools." He stressed that the key to success in U.S. schools was organizational talent and the early promotion of democracy, which he believed to be generally lacking in Mexico, especially in the rural school system. Rolland sent reports back to Carranza suggesting that his future administration should model its rural education schema on the system in Wisconsin; its "vocational nature" would best suit Mexico. Echoing the philosopher and educator John Dewey's call for hands-on learning, he recommended that boys and girls be taught physical skills: "carpentry, blacksmithing, road construction, house building, and various other useful trades."[12]

Rolland's efforts also made an impression on Americans. Supporters of Pres. Woodrow Wilson in the U.S. Congress praised Rolland's endeavors.[13] Newspapers pointed to "the commission to study the free school systems of the United States," headed by Rolland, as exhibiting the progressive nature of the Constitutionalists and the wisdom of Wilson not embarking on an invasion of Mexico.[14] Rolland was making a name for himself.

Rolland's other task was to study U.S. municipal systems. He particularly focused on hygiene and finding a way to end Mexico's *casas de vecindad*, which were multifamily buildings, usually old colonial houses that had been divided up into different units. He described them as dens of tuberculosis owned by the wealthy, who were "occupied only in collecting rents." Rolland, influenced by studies of U.S. public health trends, argued that Mexicans needed more "pure water, air, and light." He was also motivated by writings about Glasgow, Scotland. For much of the nineteenth century Glasgow had been a gray city, choked with smokestacks and full of impoverished workers. By the end of the century, however, it had become an example of progressive reform. Its municipal government had torn down slums, built new public housing, taken over certain utilities, and reduced the price of gas. The city government also took control of the local streetcar company and expanded its services.[15]

Rolland undertook his study of municipalities during a period of progressive demands for reform in the United States. U.S. progressives had pushed for the right of citizens to bypass their state legislatures by proposing statutes through popular referendums. They also pushed for increased sanitation and the ability to recall corrupt officials. Movements in the United States had in turn been influenced by town-planning trends in Britain.[16] Rolland interacted with U.S. progressives and socialists who criticized poverty, corruption, poor working conditions, and corporate monopolies.

Rolland's investigations are a clear example of how Mexican intellectuals, much like those in the United States, were influenced by a larger global reform movement. Rolland's desires for Mexico dovetailed with the idealistic and paternalistic movements initiated by progressives in the English-speaking world, in Germany, and in France. To Rolland and many other Constitutionalist leaders, the revolution would be exemplary in carrying out progressive polices. It was a progressive revolution.[17]

Although Rolland was convinced that a Mexico built on a U.S.-style education system would be a positive change, he never articulated a clear plan of how these changes could be reasonably implemented. In mid-1914 Mexico had no coherent government and consisted of millions of people who spoke different languages and who lived in vastly different types of settlements. Rolland also never entertained the idea that there might be serious contention over these foreign programs. He was confident in his abilities, but that confidence blinded him to the realities of life in most Mexican communities. This would not be the only instance in which this type of thinking would prove unrealistic and even harmful. Rolland was attempting to construct a new order out of his imagination and his experiences abroad during a period of tumultuous change.

Carranza and Rolland were prideful nationalists who stood firmly against U.S. intervention into Latin American affairs, but an appreciation of U.S. reform movements is evident in much of the Constitutionalists' planning. American organizational skills, democracy, economic growth, and power impressed them. Many Constitutionalist intellectuals believed that models for better economics, education, municipal rule, and agrarian practices were to be found in the United States, Europe, and even faraway New Zealand. In Rolland's opinion these countries' governments were generally better skilled and organized than Mexico's.

Rolland increasingly promoted himself as a socialist, specifically as a "scientific" socialist, or someone who applied science to shaping and improving social conditions. "Socialist" had become a buzzword among Constitutionalists. They used the label to differentiate themselves from their enemies, whom Carrancistas often lumped together as reactionaries. In reality this "socialism" represented a wide spectrum of thought unbound from the writings of the most stalwart proponents of socialism. Most of these newly proclaimed socialists' views fit somewhere on the ideological spectrum between laissez-faire capitalism and statism.

Originating out of the growth of the social sciences more generally, this moniker of "socialist" was in part a legacy of Comtean positivism, which was a philosophical doctrine followed by a number of Díaz's advisors during the Porfirian era. To proponents of the new social sciences, societal ills would be cured through prescribed, scientific principles and efforts, which had been developed in their most complete form in the powerful centers of Western civilization. Rolland never publicly acknowledged that he shared these ideals with his predecessors and former teachers—the *científicos*—whom he now condemned. Unlike many of them, Rolland was not a social Darwinist. He called for a more compassionate science. Rolland and many of his colleagues nonetheless saw the revolution as a war to bring science, modernity, and reform to Mexico while ridding the country of an unjust and backward colonial legacy—a vision not drastically different from that of his technocratic forebears.

The Constitutionalist Split, the Rise of Woodrow Wilson, and Rolland's Shifting Role

Back in Mexico the Constitutionalist forces had made headway in defeating Huerta's soldiers. Villa's army, the Division of the North, had permeated central Mexico in March. That army had grown to massive proportions. Zapatistas menaced the mountains south of Mexico City. Carranza's Army of the Northwest made slow but methodical progress along the Pacific coast under the leadership of Álvaro Obregón. Constitutionalist forces under Gen. Pablo González remained entrenched in northeastern and north-central Mexico. U.S. president Woodrow Wilson, who had been elected in 1912, also began to shift his support toward the Constitutionalists.

Wilson's steps to help oust Huerta were not always appreciated by the Constitutionalist leadership. On the morning of April 21, 1914, the U.S. Navy invaded Veracruz, Mexico's most important port city. After encountering resistance from Mexican naval cadets and local citizens, U.S. forces had taken control of Veracruz by April 24. Six hundred Mexicans died in the fighting. Determined that Huerta needed to go, Wilson decided on the occupation after the arrest of Americans who had entered off-limits areas of the nearby port city of Tampico and the gathering of information that there was a large shipment of arms from Europe headed to Huerta via Veracruz. The occupation made things difficult for Huerta, but it allowed him to gain support because it increased anti-American sentiment.[18] It also placed the Constitutionalists in an awkward position. Carranza could not afford to look like a puppet or that he supported a U.S. invasion. Villa and Carranza, as rumors correctly told, increasingly distrusted each other. Villa, contrary to Carranza and in an attempt to win U.S. favor, did not condemn the invasion. Mexico's revolution-turned–civil war increasingly looked like it might get even messier. Mexico and the United States were on the verge of full-blown war.[19] The governments of Argentina, Brazil, and Chile reached out to the Wilson administration, suggesting that they could moderate a cessation of hostilities in Mexico, help remove Huerta, and assist in the establishment of a compromise government. The meeting to discuss the peace deal was to take place at Niagara Falls, Ontario, and became known as the ABC Commission. A number of U.S. newspapers cited Rolland as first presenting Carranza's response that "'unconditional surrender of Huerta or of whatever other administration claims to derive its supposed authority from him or elements which maintained him in power' was the sole offer [Carranza] could accept."[20]

Rolland became a more prominent figure in providing information, in addition to gathering information, as the relationship between Villa and Carranza soured. Although the final split between the two military chieftains would not occur until October, Carranza's spies had been reporting since spring that Francisco Urquidi would likely go with Villa if there was a split.[21]

Knowing this information, Carranza turned to Rolland, who preferred the First Chief's calculated organization and more urban perspective to Villa's charismatic unpredictability and cowboy army. Carranza called on Urquidi in June, asking to meet with him in Monterrey, Nuevo León, and Rolland was to continue as the interim commercial agent in New York.[22] Rolland's stint as the main Constitutionalist representative in New York began as Huerta prepared for his impending resignation and exile. Huerta's military positions became exceptionally weak in June and early July, and his government fell apart. A number of its members stated their intention to flee to the United States. Rolland acted as the proverbial pit bull, using his press connections to attack the soon-to-be arrivals as "hypocrites," "cowards," and "robbers."[23]

Huerta resigned on July 16 and fled into exile. He left his friend Francisco Carvajal as provisional president until power was transferred to Carranza. Rolland referred to Carvajal as a corrupt *científico* and one of Huerta's strongest supporters. Constitutionalist agents had begun to use the term *científico* more and more as a label for anyone they deemed reactionary. The Constitutionalists refused to recognize Carvajal. They would not make the same mistake as Madero. They would not include prominent members of the old order in any new government.[24]

The Mexican Bureau of Information

During this time Rolland, along with Carlo di Fornaro, established the Mexican Bureau of Information to "distribute articles to magazines" and to produce a weekly bulletin called the *Mexican Letter*. They sent these reports to "500 newspapers in the republic [the United States], infiltrating the entire expanse of the nation, little by little, informing about what the Mexican nation represented, the ultimate significance of the Revolution, and the orientation we proposed for reconstruction."[25] In other words they established a propaganda mill.

Urquidi returned to New York, apparently with Carranza's blessing, in mid-August. Urquidi was surprised to find Rolland

and di Fornaro operating a press operation. Urquidi called their work "inconvenient and unpatriotic."[26] The real issue, however, was not their loyalties but that Rolland was supplanting Urquidi in his position. It is unclear if Rolland initially had permission from Carranza for these activities, but contrary to Urquidi's complaints the Constitutionalist leadership found nothing unpatriotic about Rolland's work. As it became clear that Urquidi backed Villa and that Rolland supported Carranza, the First Chief provided Rolland more official support. The move paid off. Newspapers across the United States began publishing Rolland's writings.[27]

Rolland and di Fornaro, for example, produced a number of propaganda pieces and sent them to individuals, organizations, and newspapers as the *Mexican Letter*. These writings verbally mauled Villa following the formal split of Villa and Carranza in October after the Convention of Aguascalientes. Starting in mid-October in the city of Aguascalientes, the military leaders of the revolution met, supposedly to hammer out their disagreements and to establish the foundations of post-Huerta Mexico. For a month the supporters of Carranza, Obregón, Villa, and Zapata debated, but the meeting resulted in the opposite of its proclaimed intentions. Instead of unifying the fracturing forces, it exacerbated differences and sparked a new, even bloodier round of civil war. The convention created two distinct camps—retaining their old name, the Constitutionalists, led by Carranza, Obregón, and Gonzáles, and the Conventionalists, nominally headed by convention-elected president Eulalio Gutiérrez Ortiz but really dominated by Villa and Zapata. The Conventionalists took the capital and the Constitutionalists fled to Veracruz, where U.S. forces were preparing to withdraw.

In response to the Constitutionalists-Conventionalists split, the Mexican Bureau of Information released a sixteen-page, double-columned pamphlet titled *"Red Papers" of Mexico: An Exposé of the Great Cientifico [sic] Conspiracy to Eliminate Don Venustiano Carranza*. It was a mix of documents and propaganda aimed at depicting Villa and the Conventionalists negatively and showing the Constitutionalists as morally superior. Di Fornaro and Rolland connected the Conventionalists to Díaz, Huerta, and even Madero exiles headquartered in U.S. and European cities who were eager to weaken Carranza and to rise to power through manipulating Villa. The *"Red Papers"* portrayed Villa as "fiery tempered, illiterate and politically inexperienced" and as an "impressionable general" whose vanity "was tickled." Constitutionalist propagandists painted Zapata in a similar light. Only two months earlier Rolland had called Zapata a "patriot" and a "leader of the people."[28] But now, according to Rolland and his colleagues, Zapata, though he had been a passionate promoter of social justice, had been duped by power-hungry, manipulative advisors. Di Fornaro and Rolland then used cherry-picked telegrams and letters to argue that the Convention of Aguascalientes had been hijacked and that Carranza was trying to salvage the revolution from militarists and reactionary infiltrators.[29] Like most decent propaganda, it was part truth, part exaggeration, and part creative storytelling.

A More Direct Role in Shaping Mexico

Rolland, however, was not content with being a propagandist; he wanted to shape the revolution more concretely. After all, he was a talented engineer and Carranza had tasked him with planning the reconstruction of Mexico. The Constitutionalists-Conventionalists split also made his skills more important to Carranza's operations. With Faustian zest Rolland used his writing and engineering skills to promote grand ideas and to rebuild Mexico to match his mental image of it.

On October 6, prior to the Convention of Aguascalientes, Carranza had already made Rolland the *oficial mayor* of communications.[30] Rolland continued to run the Mexican Bureau of Information, but he now helped Carranza maintain and build the infrastructure necessary to secure Constitutionalist control of the government and economy. Rolland was also elected to the board of the National Railways, recently renamed the Constitutionalist Railways, putting him in a greater position to fulfill his longtime goal of more completely nationalizing Mexican railroads.[31] E. N. Brown resigned.[32] Although the Constitutionalists failed to raise the funds needed to buy out all foreign stockholders, they did increase Mexican ownership, and Mexicans took over the main leadership positions. Rolland would not remain on the board long, but he later worked on railroad development in other capacities.

Not surprisingly, in light of his desire to reshape Mexico, Rolland had also been working on what he and many other Mexican intellectuals considered the country's greatest problem—agrarian reform. Rolland genuinely believed that the "agrarian question" was the root cause of the revolution. If Mexico was to find peace and to build a stable foundation to create a more just, equitable, and modern society, it needed to address the vast inequality in land distribution. Rolland wrote essays on the topic, drew up draft legislation, and presented his ideas to public audiences.

His work on agrarian reform has gone unnoticed by historians. They have focused on other architects of Mexico's revolutionary land policies, especially those of Andrés Molina Enríquez, who wrote *The Grand National Problems* (1909)—a strong critique of agrarian laws and practices in Mexico, especially during the Díaz government—and Luis Cabrera, who was a lawyer and friend of Molina Enríquez. But Rolland's work didn't go unnoticed among the Constitutionalist leadership. Although Cabrera drafted Carranza's famous Agrarian Decree of January 6, 1915, Rolland had presented a more radical plan than Cabrera's, a plan that was widely debated. Rolland's ideas would go on to influence experiments in land redistribution in Yucatán, Mexico's "laboratory of the revolution" in 1916 and 1917, though they would never be fully implemented.

Rolland's writings on agrarian issues provide a window into the schools of thought that motivated him. There is no doubt that Molina Enríquez's publications affected Rolland's take on agrarian reform. There are too many similarities in their works to suggest otherwise. Molina Enríquez had argued that land had become unequally divided within Mexico, with too much of it in the hands of the Catholic Church and wealthy landowners. Molina Enríquez—who had been discredited in the eyes of many Constitutionalists for not supporting Madero, for siding with Huerta, and then for joining the Conventionalists—argued that land taken unjustly should be returned and that the president should oversee the redistribution of land, placing it into the hands of communities and, even more importantly, into the hands of small, private owners.[33]

Molina Enríquez's writing contained errors, certain truths, and vague statements that proved problematic. He exaggerated the Church's power, which had declined considerably since the 1870s. But many revolutionaries, including Rolland, clung to this notion. Molina Enríquez's claim about *hacendado* landholdings was, however, irrefutable. There were a number of small ranches and farms in Mexico, but much of Mexico's fertile land was held by a small class of wealthy families. The most problematic remedy to arise out of policies influenced by Molina Enríquez was addressing his notion of lands taken "unjustly." The term insinuated something beyond unlawful; after all, many *hacendados*, railroad owners, and surveyors had lawfully taken lands long held by indigenous communities. Deciding on what was just and unjust proved to be, not surprisingly, an opaque and highly contested process.

Rolland had also been influenced by his time as an engineering student and young professor during the Porfirio Díaz administration. A number of agrarian technocrats and intellectuals had pushed for changes during the economic hardship and poor crop yields of the early 1900s. Engineers, including Rolland's mentors, were prominent in these debates. A number of Díaz's ministries had pushed for increased public and private investment in modernizing farm methods and equipment, increasing irrigation, and reducing the land controlled by large haciendas in favor of small to medium-size private farms.[34]

While in New York, Rolland published in September 1914 his first essay on the agrarian issue—*Distribución de las tierras: Estudio sobre Nueva Zeelandia, utilidad de la lección para Méx-*

ico (Distribution of lands: A study about New Zealand used as a lesson for Mexico). Rolland regurgitated some of the same arguments made by influential Mexicans, but, as the title suggests, Rolland was influenced by land reform policies in New Zealand from the early 1890s, developments that U.S. progressives discussed widely.[35] In December he would follow *Distribución de las tierras* with another work focusing on trends in New Zealand: *The Agrarian Question and Practical Means of Solving the Problem*. Rolland wanted the Constitutionalists, fresh from their victory over Huerta, to incorporate the lessons from New Zealand into Mexico's own policies for land reform.

Rolland latched onto the New Zealand model because he found similarities in New Zealand's and Mexico's histories. Mexico, like New Zealand, had a substantial indigenous population that had practiced communal forms of agriculture, it had a European colonial heritage, and it had large landholders, complicated taxes, and problems with unequal land distribution. There were also differences. Displaying the influence of nineteenth-century Mexican liberalism, Rolland argued that Mexico faced a more difficult path to peace, equality, and modernization because of a problem with religion. Rolland argued that, unlike Mexico, New Zealand had no major bloodbath because it had nothing like the massive influence the Catholic Church had on Mexico.[36]

Rolland went into detail about what he had gleaned from his study of New Zealand. In the early 1890s the New Zealand government, after being provided some autonomy by London politicians, took radical measures to make land acquisition more equitable in the face of rising immigration. According to Rolland, the state seized all land and established a perpetual rent system in which rent was based on a "reasonable interest on the intrinsic value of the property." In order to stay on the land an individual had to make certain improvements during a ten-year period. According to Rolland, "By this way capital is not needed at the beginning to acquire ownership of the land, nor must labor be pressed to pay the annual installment, which is always higher than rent. . . . By creating interests on the property which [the individual] will not willingly abandon, the State is guaranteed the presence of the colonist on his farm which national interests require that he should attend." The government also opened its own agricultural banks. Despite New Zealand's large national debt, the burden, Rolland argued, was carried easily because of the subsequent prosperity. The program was a "paternal, kind and clever policy."[37]

This essay contained some truth, but it also incorporated some key distortions. In his analysis Rolland exaggerated the nationalization of New Zealand's lands while ignoring the negative repercussions for New Zealand's native Maori population, many of whom had been pushed into alienation and hardship by contact with Western imperialism. He promoted the radical rhetoric of New Zealand reformers more than the implementation that actually took place. But for Rolland, contrary to the example provided by the architects of New Zealand's agrarian policies, the inclusion of the indigenous populations of Mexico was central to any agrarian policy in Mexico. He wanted to take properties from the landed elite and provide them to Indian families, who would receive education and become modern farmers. Rolland further contended that, like New Zealand, Mexico should nationalize mines, certain industries, and public services, including railroads, the mail, telephones, and telegraphs.[38]

Rolland's championing of New Zealand's policies also hints at the influence of an important figure who would later come to dominate much of Rolland's thought about land and political economy—Henry George. Largely a self-taught political economist, George had worked as a journalist in San Francisco and was an impassioned believer in remedying economic inequality through a single land-value tax, or "single-land tax." His first book, *Progress and Poverty* (1879), was one of the most widely read publications of the 1880s and 1890s. Praised by prominent intellectuals around the world, the book had a significant impact on the promoters of agrarian reform in New Zealand, though George's ideas were never fully realized there or anywhere else people have attempted to employ them.[39]

The civil war between the followers of Carranza and those of Villa and Zapata did little to deter Rolland. If anything, Constitutionalist leaders recognized that they needed to make their ideas about social reform better known. Villa and Zapata were popular among their own followers and with a large swath of the Mexican population more generally. Villa was part of the flotsam and jetsam of northern life—an eclectic world of miners, cowboys, sharecroppers, and drifters—and while he never got around to formulating a comprehensive plan for agrarian reform, he nevertheless understood the need for change. Zapata had already put radical land redistribution policies in place in his home state of Morelos, though his vision was more about recovery and autonomy than modernity. Despite the Constitutionalists' best efforts, Villa and Zapata hardly came off as Porfirian reactionaries. To attract and maintain support from peasants and laborers, the Constitutionalists needed to make clear that they would not bring back the inequitable policies of Díaz or the overly cautious rate of change put forth by the Madero administration. Carranza and other Constitutionalist leaders decided to hold a conference in their base port of Veracruz to hammer out, clarify, and then publish their stance on important issues.

The Veracruz Conference

Rolland, who had already been spending more time in Mexico in his new role as a top communications official, spoke often about land reform and the nationalization of public utilities. On October 23, while the Convention of Aguascalientes was under way, Rolland gave a presentation in Mexico City on his thoughts about New Zealand and the agrarian issue.[40] In Veracruz he addressed his colleagues and the public in multiple lectures. He also talked with journalists, reemphasizing a "practical nationalization of lands" from "a just and scientific standpoint." Once in power, the Constitutionalist government should, he proclaimed, confiscate all unlawfully acquired lands, expropriate other lands (paying owners a fair price), and put in place a perpetuity rent system that had a one-year tax waiver and a requirement to improve the value of the property.⁴¹ His plan offered a way to establish an equitable and far-reaching tax structure, to break down *haciendas*, and to get around vague notions of lands taken unjustly. All land would be taken and then given. The government would be the ultimate arbiter.

As Rolland exemplifies, Constitutionalist leaders varied in their visions of Mexico's future. They were by no means a monolithic force. They were not moved solely by political motivations.⁴² Many Constitutionalists were genuine in their belief that agrarian reform was the foundation upon which any revolutionary government had to be built. Rolland was correct when he stated that many Constitutionalist intellectuals were at least "partly guided by socialist motives."⁴³ These men heatedly debated land, labor, and the role of the state.

During the last week of December, Rolland had conversations with dozens of Constitutionalist leaders. He and Geraldo Murillo, better known as Dr. Atl, established the Revolutionary Confederation to push for more radical changes.⁴⁴ Other members included Rafael Zubarán Capmany, Alberto J. Pani, and the Sonoran generals Adolfo de la Huerta, Álvaro Obregón, and Salvador Alvarado. The Revolutionary Confederation's stated goals were to exchange ideas, build cohesion among Constitutionalists, and influence policy. They wanted to unify revolutionary goals and efforts among the Constitutionalists, defend the rights of the oppressed, develop legislation to secure those rights, work to bring about a quick military triumph, and collaborate with "men and women" to fight against the "old militarism" that "in the name of clericism and capitalism" had caused "all national tragedies."⁴⁵ With a growing number of adherents, they attempted to put greater pressure on Carranza, whom Rolland believed to be acting too timidly, and to shape future government policy.

Carranza himself possessed a conservative bent. He was a landowner, and his Plan of Guadalupe had intentionally excluded an agrarian reform policy. He had returned confiscated lands to U.S. business interests. But Carranza knew that land reform would have to be addressed if the Constitutionalists were to gain more support from the thousands of peasant soldiers who took up arms because they felt robbed of their lands or wanted land. And he did not disagree with the statement that land distribution had been historically unequal since colonial times. Conservative but by no means blind, he was willing to experiment with limited reform if that was the price of ensuring his government's ultimate survival and ending the war.

A number of the policies hammered out in December were incorporated into decrees issued during the following month. Most famous of these was the Agrarian Decree of January 6, 1915. The law essentially offered a conservative application of the writings of Molina Enríquez. Drafted largely by Cabrera, it proposed to nullify land contracts made after 1876. It stated that the central government made the ultimate decisions on land and resource policies, and it proposed to return *ejido* (communal) lands, whether the titles were clear or not, to pueblos that had had land "unjustly" taken from them. It also established that the executive branch would determine land redistribution policies. Carranza would form what would be called the National Agrarian Commission, and each governor, military or civilian, would choose people to lead local agrarian commissions.[46]

Many historians have been suspicious about Carranza's motives, arguing that he was carrying out a political ploy to destroy Zapata and Villa's monopoly on radical land policies in order to woo potential soldiers and gain support among the general public.[47] The skepticism is warranted even if wrongly applied to the entire Constitutionalist leadership. Carranza was fairly conservative, and more than that he was pragmatic. And despite believing that some form of land redistribution was necessary, he nevertheless did not want to alienate powerful foreign governments and business operators. To Carranza, the Constitutionalists had to walk a tightrope on the subject.

But the cynicism of later historians undermines the varied and contentious arguments about land policies within the Constitutionalist camp. The Revolutionary Confederation backed Rolland's plan. The January decree infuriated Rolland. After it was publicized, Rolland, Dr. Atl, and Zubarán complained to Cabrera and Carranza. According to one journalist, Rolland, with intense eyes, was "twisted with disappointment." He bitterly commented that the law was "too concerned with the ejido, only the ejido. The ejido is something of the past. . . . Further, there are many men in Mexico that are not part of a 'pueblo' and who also need lands. We want to talk with Cabrera to make our thoughts clear." Dr. Atl seconded Rolland. The "law is deficient, only resolving part of the question."[48] In their attempt to build a unified vision in Veracruz, the Constitutionalists began to see their latent divisions rise to the surface. These divisions would grow as the revolution progressed, causing rifts that influenced revolutionary governments, playing out within competing bureaucracies over decades.

Rolland and his allies failed to persuade Cabrera or Carranza, but that did not lay the issue to rest. Over the next couple of years Rolland worked with Gen. Salvador Alvarado to implement a more radical policy at the local level in Yucatán. There were often significant differences between Carranza and those officials who implemented Constitutionalist policies. How these differences played out often determined the final results.

Perhaps because of these differences, Carranza ordered Rolland to work on a new task. Rolland was to continue propaganda operations in the United States and to join a new commission headed by Pastor Rouaix, who was an engineer and the undersecretary of development. This new commission would study the U.S. oil industry.[49] Before his departure from Veracruz, Carranza and Cabrera ordered 5,000 pesos released to support Rolland's "propaganda, press, and information services" in the United States.[50] The Petroleum Technical Commission's specific mandate was to uncover new developments in the U.S. oil industry, examine the relationship between the Mexican government and these industries, and develop proposals for future Mexican legislation that would protect and develop the resource.[51] Rolland was just as radical about the nationalization of the oil industry as he was about land, but he found more sympathy for plans to nationalize subsoil resources.

Preparations for the commission took longer than planned. Although there were monthly reports starting in January that indicated the group was on its way to the United States, it did not embark on its journey until early May 1915. It spent some of the intervening period north of the city of Veracruz, exploring oil fields that were controlled by foreign corporations. The Petroleum Technical Commission's delay may also have resulted from the fact that Rolland, while engaging in policy debates, had been overseeing the construction of a shipping dock in Veracruz. He was also working with a public health commission on improving water and sanitation in the city's markets.[52]

Rolland continued to work on policies and press statements, too. In one article addressed to U.S. progressives, "A Trial of Socialism in Mexico," Rolland praised socialism as the pinnacle of good governance while attacking the exploitation inherent in capitalist and religious systems. Rolland went so far as to argue that Mexico was on its way to becoming its own example of the benefits of socialist reconstruction and that even leaders in the United States should take notes because, despite the United States' material progress, it still faced massive problems of capitalist exploitation. Rolland further claimed that the revolutionary chiefs under Carranza had the "most profound conviction that the socialist state be established, that the control of services of public utility should be without speculative aims, and that [there be] the creation of small interest by the reapportionment among all natives of land holdings and of the natural resources of the country." By this means the Constitutionalists would create a nation based on "peace and happiness," not "a nation prepared to kill, like Germany, or one organized mainly for material gain like the United States of America."[53]

The policy prescriptions Rolland gave, however, were his, not Carranza's. And though some colleagues agreed with some of what Rolland advocated, they often did not share his exact vision.

The future shape of the Mexican government was contested, and Rolland pushed his own agenda. During the years that followed, Rolland would not only expand his endeavors to influence Mexicans and Americans, he would align with Alvarado, who took charge of Yucatán in 1915 in an attempt to make his vision a reality.

4

Back to the Periphery

MODESTO ROLLAND AND THE rest of the Petroleum Technical Commission set off on the U.S. Ward Line steamer *Morro Castle* from the Gulf of Mexico port of Tampico in early May 1915. The ship's two large smokestacks churned out exhaust as the liner smashed through the waves. Rolland was thirty-four, slightly older than the other five members, excepting Pastor Rouaix. While boarding the boat, the commission stopped for a photo op. Their suits were well tailored and clean. Their mustaches were long, slightly twisted up at the ends, and well manicured. Some of them had already broken out their straw boater hats. They were going on a whirlwind adventure of sorts. They were on a mission to study the petroleum industry, traveling first to Cuba and then across the United States, from New York to Los Angeles, stopping in cities big and small in between.[1]

Over the next five years, at the height of Constitutionalist power in Mexico, Rolland would go on to do much more than delve into the particulars of the oil industry, though petroleum development would remain a prominent theme in his work. Indeed he would take some of his newly gained petroleum knowledge and apply it immediately to his goals of developing Mexico's peripheries and expanding infrastructure and industry. During his regular visits to New York, Rolland would also catch up with Carlo di Fornaro, their press operations, and their American contacts. Rolland didn't let up on his drive to influence opinion in the United States and Mexico. He continued to promote his vision for a "socialist" Mexico, agrarian reform, and, increasingly, the economic policies of Henry George.

Rolland also somehow managed to turn serious attention once again to developing Mexico's frontier peninsulas: Yucatán and Baja California. Not only does that effort indicate his continuing belief in regional development as key to national integration, it also exhibits the growth and application of Rolland's internationally influenced progressive ideals, the continuation of allegiances made in Veracruz, and Rolland's first taste of stiff resistance to his plans from local residents and the president alike. Of course Rolland could only imagine things yet to come as he set out on the *Morro Castle* in 1915. For the time being Cuba was on the horizon and New York waited beyond.

The Petroleum Technical Commission Sets Sail

The petroleum commission made its first stop in Cuba within a day of leaving Tampico. It was a brief layover. Havana was just winding down from a widely attended boxing contest in April between juggernaut Jack Johnson and his "former Kansas cowboy" challenger, Jess Willard. Mario García Menocal, the aristocratic president of Cuba, himself an engineer with a degree from Cornell University, had visited Willard in late March, expressing "admiration for the young giant, although he had previously bet $100 on Johnson."[2] Menocal, to his disappointment, lost his bet. In a match witnessed by thirty-two thousand people, Willard became the new heavyweight boxing champion.[3] Despite the spectacle of sport, there had been unrest in Cuba. Workers' voices for rights and better conditions had become loud. The Cuban government, still adapting to its independence from Spain and subsequent subservience to the United States, had in the years leading up to Menocal's election in 1912 gone through a massive expansion in government jobs and public works. There were new roads, telegraph stations, and an increasingly bloated bureaucracy.[4]

The Petroleum Technical Commission used its brief stopover to meet with the country's minister of agriculture, who provided data about oil production and petroleum laws on the island.[5]

Considering the brevity of the stopover in Havana, the information obtained was surely limited. But what the commission did find was that petroleum regulations were pretty much nonexistent. It was clear that U.S. business interests were entrenched and that Cuba relied on the United States for petroleum, despite talk of Cuba expanding its own exploration efforts.[6] Carranza hoped to build stronger ties with Cuba and other Latin American countries, and visits like this one helped form international bonds while reinforcing the Constitutionalists' call for stronger restrictions on large U.S. corporations.

As Rolland, Rouaix, and company disembarked in New York City on May 10, ships and war were very much on the minds of many New Yorkers. World War I, fueled by a German assault across western Europe a year earlier, dragged on despite German hopes for a quick victory. Just three days before the arrival of the commission, a German U-boat had torpedoed and sunk the massive British ocean liner *Lusitania*, killing 1,198 passengers, including a number of Americans. Editorials in the city's newspapers clamored for "strict accountability" for the Germans, if not outright war against them.[7]

News of the *Lusitania*'s sinking did not slow the commission. It gathered everything it could on the petroleum industry. Its members combed bookstores and libraries for pamphlets, books, and magazines. In Washington DC its members discussed petroleum regulation and exploration with members of the Interstate Commerce Commission, U.S. Geological Survey, General Land Office, and Bureau of Mines. According to José Vázquez Schiaffino, a Mexican engineer who made the trip, one of the officials at the General Land Office told the Petroleum Technical Commission that "the [Mexican] government should not sell public lands but should instead rent on terms favorable to the nation . . . [and] governments should conserve all natural resources, renting them under certain terms, but never absolutely ceding them."[8] One of the strongest influences on the commission, very much contrary to the interests of American oil firms, appears to have been members of the U.S. government who promoted a stronger role for the government in protecting and developing resources.

The commission subsequently moved west, going through Baltimore, Pittsburgh, Chicago, Tulsa, Los Angeles, and San Francisco. Its members met with representatives from Standard Oil, the United States Asphalt Refining Company, and the California Oil and Asphalt Company.[9] The asphalt companies were of particular interest because they were an importer of Mexico's particularly viscous oil. The commission soon discovered that it was not the only body of foreigners studying the American oil business. Only a week before the Petroleum Technical Commission arrived in Independence, Kansas, a small town near the Oklahoma state line, to study Oklahoma oil fields owned by the Prairie Oil and Gas Company, the Japanese naval commander Yusuke Miyamoto had arrived in the community for the same purpose.[10] In Mexico there were U.S. and British oil companies scrambling to buy land near Tampico while regulations remained weak.[11] Perhaps they sensed that the pickings would not be so easy in the near future.

The trip proved immensely important. The commission determined that U.S. petroleum companies were booming and that they faced stiffer regulations at home than they did in Mexico. The commission called on Carranza to enact more regulations. In a suggestion that had immense consequences, their report also discussed the possibility of nationalizing subsoil resources, a return to Spanish colonial policy but now with modern energy-sector needs in mind. The Petroleum Technical Commission also purchased equipment and chemicals to establish a petroleum laboratory in Veracruz and to better equip the Geological Institute in Mexico City. Early in 1916 Carranza issued a number of decrees that changed policies toward foreign oil companies; among them were changes in "taxation and regulation of petroleum production."[12] U.S. oil companies complained bitterly that the Constitutionalists were anti-American.[13] Rouaix would go on to author much of Article 27 in the Constitution of 1917, which declared all subsoil resources as state property.[14] That said, he and Carranza both preferred that most of the companies stay and that they simply comply with new fees, leases, and regulations.

The commission also left a lasting impression on Rolland. He was now certain that Mexico's modernization and independence depended on developing an oil industry controlled by Mexicans. Not only did Mexicans need to explore for undiscovered oil deposits, they needed to wrest control of the industry from foreigners and develop the equipment and infrastructure necessary to refine, store, and ship petroleum domestically and abroad.

Agrarian Reform in Yucatán

While participating in this cross-continental journey, Rolland continued his relationship with Salvador Alvarado. After a brief military campaign by Alvarado and Gen. Heriberto Jara into Yucatán, Carranza had named Alvarado governor of the southern region in February 1915 to solidify control over the area's lucrative henequen industry.[15] U.S. and Canadian businesses used henequen to make twine for binding wheat.[16] The fiber had become a hot commodity before World War I, and the war made henequen all the more valuable. Alvarado had by March established himself firmly in the state's capital, Mérida. His goal was to revolutionize Yucatecan society by destroying the hacienda order and replacing it with a more equitable, state-driven, capitalist economy, which he would oversee. Because of the distance between Mérida and Mexico City—which the Constitutionalists had recently regained—and the ongoing military conflicts with Villa and Zapata, among others, Alvarado initially possessed significant autonomy to direct a top-down reformation of Yucatecan politics, economics, and society.[17] Alvarado, knowing Rolland's "revolutionary enthusiasm," invited him to organize the state's agrarian commission and land office. During their time together in Veracruz, Alvarado had been impressed by Rolland's initiatives, and the new military governor allowed Rolland to attempt, with certain constraints, to carry out his agenda. Alvarado was also interested in Rolland's knowledge about education and the petroleum industry.

Upon returning to Mexico, Rolland and his family packed up their belongings and headed to the southern peninsula. They established themselves in Mérida, the white-walled capital of Yucatán. Spanish-descended henequen elites dominated high society, but the region's prosperity was built on human exploitation, mostly of indigenous peoples. There was a large and mostly rural Maya population working as servants and hacienda laborers. Yucatecos shared a history of bloody wars fueled by divisions between the rich and the poor, the urban and the rural, and the Maya and the non-Maya. And Yucatecos of all ethnicities distrusted outsiders, including Mexicans from the north. The land around Mérida was flat, covered in shrubs, and dotted with the henequen agave plant, also known as sisal. Like Baja California, Yucatán was connected to foreign interests and possessed weaker ties to Mexico City than central Mexico.

Alvarado placed Rolland in charge of land redistribution and revaluation.[18] Rolland assembled a staff consisting of a number of his former colleagues and students from Mexico City—"a large staff of engineers" to "destroy the haciendas," a task they took to with "great vim."[19] He hoped this was the first step toward turning poor Maya peasants into a prosperous, small-propertied class based on a single land-value tax. Land confiscated from reactionaries—*hacendados* found guilty of rebellion or of stealing lands from Indian communities—and from the Church would be managed by the state or divided into small farms worked by individual families. The commission confiscated other properties, reimbursing large landholders with funds they partially took from neighboring property holders via increased property taxes. By October 1916 Rolland was claiming that the commission had redistributed properties to approximately forty thousand families, a number that seems too large to be accurate. It provided twenty hectares, or nearly fifty acres, to each male head of household if the land was uncultivated and ten hectares if the land was already under henequen cultivation.

Rolland, Alvarado, and other members of the revolutionary entourage wanted to elevate the "Indian race" and transform it into a new, improved civilization.[20] Like many of his peers, Rolland used the term *indio*, or Indian, loosely, often in reference to all poor, brown-skinned, rural Mexicans, though many of the communities Rolland worked with in Yucatán had a mostly Maya population. Mexican intellectuals had evolved their perception of the peoples they labeled Indian over previous decades as global intellectual trends shifted. In the early Porfirian era the view that Indians were members of a biological race inherently inferior to people of European ancestry was common. By the end of the 1800s, however, people like Secretary of Education Justo Sierra had argued that difficult conditions in indigenous communities had nothing to do with racial inferiority but instead had to do with lack of education, as well as other social and economic conditions.[21] Revolutionary elites such as Rolland built on this latter trend, arguing that Indians were capable learners but people who needed to be educated about how to be productive Mexican citizens. Revolutionary officials undertook a crusade of "celebrating" Mexico's indigenousness while attempting to assimilate Indians into a greater Mexican identity, a campaign that became known as *indigenismo*.[22]

To Rolland, the best way to improve the lives of Indians was agrarian reform, especially reform done in a way that put into practice the ideas of Henry George. Rolland had become interested in George sometime around 1913, but this influence had largely been indirect, coming from U.S. progressives and pamphlets about New Zealand. This changed in the fall of 1915, when Rolland became more directly involved in the "Georgist" movement. Rolland indicated that he had returned to New York yet again to continue his work as a propagandist. With the help of di Fornaro, Carranza, Alvarado, and a couple of U.S. investors, Rolland organized the Latin-American News Association. While in New York City he continued to build contacts with "partisans of the Single Tax." He devoured books on the topic. Prominent progressives, including Lincoln Steffens and George L. Record, advocated for George's single tax, which they argued small property owners could more easily bear while still providing local municipalities and the central government sufficient revenue. Record and other Georgists "envisioned Mexico as a testing ground and laboratory for the pursuit of political and economic democracy [via the single tax]."[23] As Record put it, "In the case of a peon on his farm, it would be a payment that he could easily afford to make; in the case of an oil well or mine [or henequen hacienda], it would be a big corporate payment for the privilege of utilizing such valued natural resources and would leave an abundant return for the capital and labor invested in the enterprise."[24] In August 1916 Rolland attended a single-tax convention held in Niagara Falls, New York. He was "deeply moved by the sight of the young men from Philadelphia, who yearned to form a Georgist political party in spite of the disapproval of their elders who did not desire to arouse against the Single Tax opposition of the Republican and Democratic parties." Rolland continued, noting that since this event his "spiritual thirst for a correct principle of true social justice" had been satisfied.[25] Until his death many years later Rolland remained an ardent disciple of George.

It should be noted that at this time in Mexico the tax structure was weak. Governments since Mexican independence had been financed by a patchwork of fees, local taxes, and import and export tariffs. Carranza funded the Constitutionalist government and army largely through controlling Mexico's ports and the regions that produced profitable exports and, in turn, certain fees. There was plenty of room for creating a more solid tax system in Mexico, but similar to the experience of Rolland's U.S. counterparts, Rolland found that most political leaders in Mexico were not willing to adopt such unconventional measures. In general they continued to focus on forced loans, import and export fees, and indirect taxes and to discuss more classically liberal concepts. Of course no matter what income-generating system the state put in place, its officials had to have enough control to enforce it, which had been another difficulty for Mexican governments.[26]

Georgist prescriptions were present in the land policies under Alvarado, even if Rolland exaggerated his claims that Yucatán was on the path to becoming a Georgist utopia. The land that surveyors divided up, twenty hectares, was close to the forty-acre allotments that George had recommended.[27] The Alvarado administration also held the title to land redistributed to individual families for two years. It was only handed over to the head of household if there was proof of improvement. *Campesinos* who failed to productively work their land could face a stiff tax. By these means unproductive farmers would be forced off their land and productive owners would eventually reap larger profits because the land tax would be based only on the value of the land itself, not on the products of labor or capital.[28] The goal was to provide positive and negative motivators to get local residents to develop productive agriculture and to start industries. Of course many Maya residents saw the land as theirs to begin with. But to Rolland, like George, new farmers had to prove their capability and willingness to produce. The Alvarado administration also established an agricultural school and hired experts to help increase production. Under the influence of Rolland, many of the newly trained agronomists promoted Georgist prescriptions.[29] The school and Rolland's ideas for agrarian reform received widespread attention, largely a result of Rolland's constant promotion across Mexico and the United States.

Alvarado respected the views of George and Rolland, but he never attempted to put into place a strictly Georgist system. Alvarado worried that such measures would threaten economic stability and the wealth funneling into the fragile revolutionary government through the henequen industry. Demand for the fiber from that plant was huge, and the North American wheat crops that would be bound with henequen-derived twine had been doing well. It was unclear if the U.S. government would tolerate anything that could possibly jeopardize the twine industry and, in turn, American wheat harvests. Alvarado's power also relied on a cross-class populist alliance. In addition to his cadre of advisors from northern Mexico, Alvarado had incorporated Yucatecans into his bureaucracy, including some elites. Nationalizing and redistributing all land would make too many enemies. Alvarado did, however, agree with Rolland's desire to make land use more equitable, and he therefore allowed Rolland to revalue and redistribute large parcels of land, even without clear approval from Carranza.

Rolland's actions did not sit well with Carranza. The First Chief was wary of Rolland's ideas and the fast pace of change under way on the peninsula. In late 1916 Rolland reached out to Carranza in an attempt to show the success of agrarian reform in Yucatán, hoping it would serve as a model for the rest of Mexico.[30] Not only did Carranza reject Rolland's ideas, he ordered Alvarado to halt Rolland's work. Upset, the First Chief wrote to Alvarado that the Agrarian Decree of January 6, 1915, had referred exclusively "to the redistribution of ejido lands taken from villages and did not include returning lands that are not part of the ejidos, which constitutes another aspect of the agrarian problem in which the chief executive has not enacted legislation."[31] Carranza further rejected a proposal that would have allowed Yucatecan officials to base agrarian reform on their own state laws. An infuriated Carranza completely halted the distribution of land by Constitutionalist governors and local agrarian commissions across Mexico.[32] Much to their frustration, Rolland and Alvarado discovered that despite distance from Mexico City they had to compromise their plans for land distribution and put in place those decreed by Carranza.

Even when Alvarado and Rolland were able to enact their plans, things did not always go as intended. The owners of large estates fought attempts to take what they perceived as their property. Often local residents of lesser means who received land grants did not use them as Rolland had demanded. Some families used their land to harvest and sell firewood and make charcoal instead of growing crops. Many Maya were more interested in returning to their own practices than becoming individual, small-scale commercial farmers. At least one engineer who helped carry out reforms quit in frustration over indigenous communities who failed to follow the program design and, according to him, were reducing the land to waste.[33] Maybe there was some truth to this accusation, but this engineer and Rolland failed to appreciate the intricacies of local land use and customs, as well as the problems associated with enforcing top-down programs on people to whom the language, culture, and the goals of state agents were foreign. Rolland often spoke derogatorily of locals when talking with his associates. And though many of the officials who worked under Rolland appreciated his passion and tireless work ethic, he was loud. He shouted orders at them, sometimes rubbing them the wrong way.[34]

Beyond Agrarian Reform

Rolland's work in Yucatecan society expanded beyond agrarian reform. He became involved with the feminist movement and helped expand Yucatán's industries and infrastructure. Rolland helped craft and promote Alvarado's new labor laws. Rolland had built a working relationship with the American Federation of Labor and with its leader, Samuel Gompers, forging transnational alliances. Building on trends in the United States, the Alvarado administration provided paid maternal leave to women for the thirty days before and after a child's birth and established requirements for sanitary rooms for breast-feeding in places of employment, along with paid time to nurse. Alvarado's government established an eight-hour work day and forty-hour work week. It also created compulsory labor-capitalist arbitration laws and legislation prohibiting child labor in sweatshops. These policies would influence similar guarantees written into the 1917 national constitution.[35]

Rolland and Alvarado, similar to Gompers and most U.S. progressives, were ardent promoters of improving the lives of workers, but they were not radicals who promoted the notion of workers overthrowing the capitalist system. The general and engineer were members of a progressive, increasingly global, "socialist" bourgeoisie. Rolland argued for arbitration over striking, which he believed should always be a last resort for workers.

George's philosophies had a strong influence on Rolland and, to a lesser extent, on Alvarado. George had considered himself a genuine friend of workers, but he cared little for strikes. As he wrote in *Progress and Poverty*, "The struggle of endurance involved in a strike is, really, what it has often been compared to—a war; and, like all war, it lessens wealth. And the organization for it must, like the organization for war, be tyrannical. As even the man who would fight for freedom, must, when he enters an army, give up his personal freedom and become a mere part in a great machine, so must it be with workmen who organize for a strike. These combinations are, therefore, necessarily destructive of the very things which workmen seek to gain through them—wealth and freedom."[36]

A faithful disciple of George, Rolland promoted order and growth over actions that threatened destabilization. Alvarado, dependent on the success of the henequen industry, also possessed little tolerance for strikes or chaos, though he drew more of his inspiration from the British self-help guru Samuel Smiles, who supported a regulated form of capitalism and argued that poverty was largely a cause of worker immorality and lack of thrift.[37]

Rolland was an important official in Alvarado's Compañía de Fomento del Sureste, or Development Company of the Southeast, a state-run enterprise. The company was supposed to be founded on 100 million pesos in capital. It was to function for one hundred years, focusing on infrastructure, resource development, and trade. Half of the start-up capital was supposed to come from the Carranza administration. However, the company never received the outside funds, and Carranza may not have even agreed to the arrangement. The details remain unclear. Alvarado had difficulty obtaining the funds on his end, too, managing to raise only 5 million pesos.[38]

Despite the huge setbacks, the company, in which Rolland became a manager, was remarkably ambitious. It set out to connect Yucatán, Campeche, Tabasco, Chiapas, and the Territory of Quintana Roo. The main goal was to build a larger regional market and to connect Yucatán to the Tehuantepec Railway and beyond, to Mexico City. Rolland argued that food and resources from Tehuantepec would fuel Yucatecan industrialization. The connection to the railway across Tehuantepec would provide access to trans-Pacific trade and allow for the importation of oil from the northern Gulf of Mexico. The company established a steamship service, hoping to create its own merchant marine. Rolland headed an oil exploratory group and built works at the port of Progreso, including a petroleum storage facility. He hoped to find oil to help fuel domestic agriculture and industrial production and, if sufficient, to export to Asia, the United States, and Europe. Rolland was unsuccessful in his search, but Petróleos Mexicanos (PEMEX), the government-owned company created in 1938, later used the records he left behind in a successful attempt to reach oil beneath Yucatán's limestone substratum.[39]

Rolland's influence also extended to women's rights. Rolland helped organize the state's first two feminist congresses, the first of which took place in Mérida in 1916.[40] In Rolland's words the congress consisted of "more than two thousand women of the middle class, who, only a short while ago, were enslaved by all kinds of prejudice, and at this meeting discussed enthusiastically Education, Religion and Physiology, showing in the most unsuspected manner, the strength of feminine intellect, as well as its moral power by the side of man, to direct the future of the Mexican family."[41] The congresses were organized to discuss women's rights and the political incorporation of women in order to create a modern and just world. The conference was a genuine accomplishment. But as Rolland's writings exhibit, a patriarchal and classist aura permeated the event and the male revolutionaries' promotion of women's rights more generally. The topics were "traditionally" feminine. Women worked "at the side" of men.[42] The women participating in the feminist congresses were middle class, even elite, and mostly non-Indian.

Rolland wrote about women in Mexico more generally, too. He contended that "the history of Women in the Latin-American countries is a page full of sadness" but that the Mexican Revolution accelerated the "vindication of woman in America," providing the "most typical case of woman's evolution from a beast of burden into a human being." Rolland further heaped accolades upon the women of Mexico for participating in the revolution as protesters, as providers and caretakers of male soldiers, and as soldiers themselves. He applauded the feminist movement in Mexico and the United States. He had special praise for "women teachers and, in general[,] women of the middle class, [who] are taking active participation in public affairs, and occupy positions in all public and private offices."[43] He supported, at least in the press, women's work and economic independence. He approved of the Law of Divorce, put in place by Carranza, which made divorce more accessible.

This profeminist stance, however, begs the question: How did Rolland reconcile this with the way he treated his own wife, Virginia Garza de Rolland? He had shamed her into withdrawing a claim for divorce in 1908. She had been an award-winning student who had shown great promise as a reformer and teacher. Despite living in Mérida with Rolland and their children, there is no evidence that Garza de Rolland played any part in organizing the feminist congress or in educational reform. While Rolland worked on redeeming women and peasants in Yucatán, Garza de Rolland cared for their four children. She catered to his constant moving, politicking, and projects. Unfortunately for scholars, Garza de Rolland left little in the way of a historical record and Rolland rarely wrote about her. If she participated in any meaningful way in public affairs, it went undocumented. There is no way of knowing what she thought and felt about Rolland's work or her place within their family except that she often felt ignored. Since the death of her father, she had become more economically dependent on Rolland and she was under pressure to stand firmly by his side, as women were still told to do despite feminist rhetoric.

Even if Alvarado, Rolland, and other revolutionary movers and shakers in Yucatán were sincere about their desire to improve the lives of women, other motivations lingered just beneath the surface. Religion and politics were prominent concerns of these men. They hoped to incorporate women as allies in an effort to dismantle the power of the Catholic Church and to keep Alvarado's clique in political power. But Alvarado argued that it was the women who needed allies to fight against the oligarchic men who had dominated the peninsula's politics and economics. They needed to be educated out of their ignorant stupor, and it was revolutionary men who would liberate them. These men considered women to possess little agency of their own and that they were as a group inherently gullible.

Women of course were not passive. They actively participated in the reforms under way, but the results of their actions were not what Alvarado and Rolland had bargained for. Rural peasant women protested vociferously at times against anticlerical measures and the destruction of churches.[44] When women of the feminist congress voted on whether to support immediate suffrage, 633 of the 700 attendees voted no, citing conservative arguments that women should focus on the domestic sphere or on a more moderate, gradual process of incorporating women socially and politically into the male-dominated arena.[45]

Alvarado's administration caused resentment in other ways, too. Despite resisting some of Carranza's decrees, Alvarado's government enacted its own form of top-down policies in the name of progress and Constitutionalism. Alvarado married a local woman and built local alliances, but he and most of his advisors were not from the region. Alvarado promoted workers' rights, including the right to strike, but he pushed those same workers to choose arbitration over striking. He distributed lands to many peasants but ultimately defended some large landholders in order to maintain peace and the lucrative henequen industry. The Alvarado administration, as one historian has suggested, came off to some locals as self-righteous and arrogant.[46] For example, Rolland told the *New York Times* that "there was nothing in Yucatán's old world that anybody wanted to keep, nothing of sufficient merit . . . to waste time arguing about."⁴⁷ The Alvarado government increased the number of schools, the literacy rate, and labor policies while working to distribute land to the landless and endorsing women's rights. For many people Alvarado became a savior. But to others he and his aides were arrogant outsiders who didn't respect Yucatecan culture.

Overall, Alvarado's reforms were wide-reaching and relatively successful. In the words of the historian Gilbert Joseph, Yucatán "came to be regarded by the rest of the republic as a pace setter, a social laboratory for the Revolution where bold experiments in practical organization, state intervention in the economy, and labor and educational reform were carried out."⁴⁸ The process of land redistribution across other parts of Mexico had been much more uneven. It had progressed in slow, confused spurts in parts of Veracruz. In Tamaulipas the governor, César López de Lara, an opponent of land reform, disbanded the local agrarian commission. When the Carranza administration forced him to reinstate it, he placed in its leadership friends who shared his conservative convictions.⁴⁹ Meanwhile the Alvarado administration was redistributing lands at a pace too fast for Carranza. It carried out the abolition of the debt-labor system and dramatically increased the number of secular schools. Alvarado, like a good progressive, prohibited the sale of alcohol, ended cockfights and bullfights, and promoted the establishment of baseball teams, changes Rolland publicly advocated.⁵⁰ Frank Tannenbaum, among the first scholars to write about the revolution, wrote that Alvarado's coming to Yucatán was "like a cyclone that destroyed feudalism rooted deep in the soil. . . . He, perhaps more than any other Mexican who took part in the Revolution, attempted to formulate its program."⁵¹ A more nuanced assessment would be that Alvarado's officials destroyed the legal underpinnings of debt peonage but had a more limited impact on the conditions and attitudes that had fueled abuses. Nonetheless, for a short period of time Yucatán was paraded as a shining example of the potential success of the Mexican Revolution.

Rolland has for the most part been left out of this story despite being one of the important architects of the new, progressive Mexico. Alvarado was not the mastermind behind all of his administration's programs. It was the people around him, a small army of bureaucrats who developed and carried out the programs. Rolland's fingerprints are all over many of the most important projects. Historians have often credited Cabrera and Molina Enríquez as the dominate influences on Mexican land reform, including in Yucatán, thus underemphasizing Rolland's influence. Rolland had written his own essays and policy proposals about agrarian reform and had been formulating plans for education, land, infrastructure, and petroleum development before ever coming to Yucatán. Rolland oversaw the construction of port facilities in Progreso, explored for oil, and developed plans to link Yucatán to Tehuantepec and the nation's capital. He would build on these plans for decades, becoming one of Mexico's most prominent developers of Tehuantepec, the frontiers, and ports. Alvarado respected Rolland's thoughts and, more importantly, his ability to turn those ideas into physical reality. Rolland was provided the opportunity to help reorganize the economy and infrastructure of one of Mexico's most remote but wealthiest regions, and in turn Alvarado got to take the credit, expanding his political influence.

Baja California

Somewhat astonishingly, Rolland found time to work on issues outside of Yucatán, managing to return some attention to the other peninsula in Mexico, Baja California. In 1916 he typed an essay entitled "Problema de la Baja California." Rolland reiterated many of the points he had made back in 1911, though his writing had become more aggressive and loudly nationalist. Rolland lambasted U.S. imperialism in Latin America from the U.S.-Mexican War (1846–48) to U.S. invasions under way in the 1910s in Central America and the Caribbean. In the United States he worked with U.S. and Mexican intellectuals to gently influence the U.S. government to pursue a policy of nonintervention in Mexico. Rolland's writings in Mexico were less subtle. Rolland also reiterated his critique of inequitable taxes, the importance of municipal autonomy, and the need to defend Mexico's resources.

Rolland's not particularly novel though no less unfounded argument was that Baja California remained dangerously close to being usurped by the United States. American newspaper owners and members of Congress had called for the annexation of the region. Communication between the peninsula and Mexico City remained dangerously weak despite the construction of a handful of radiotelegraph stations. Railways in the regions favored the United States, not Mexico. The argument made among U.S. supporters of annexation was that the peninsula was more geographically connected to the U.S. state of California than to mainland Mexico, which remained physically tied to the peninsula only by a small sliver of the Sonoran Desert.[52]

To counter the long-standing U.S. threat, Rolland offered a unique solution: the Carrancista government should build a canal across northern Baja California, from the mouth of the Colorado River at the Sea of Cortez to the Pacific near Tijuana, turning Baja California into an island. In addition to creating a new physical boundary the waterway would stimulate local development in the region, better connecting mainland Mexico to Baja California and trade in the Pacific. Rolland specifically indicated that the states of Sonora, Sinaloa, Durango, and Chihuahua would see significant economic gain from easier access to the Pacific and the markets of Mexicali, Tijuana, and the western United States. These Mexican states, according to Rolland, should raise money through loans, which in combination with national bond sales would fund the project. To sweeten the concept, Rolland provided an account of the agricultural, marine, and mineral resources of the peninsula, especially in Mexico's valleys bordering the United States, which produced a wide variety of fruits and cotton. He also promoted the exploitation of orchilla weed, which was used to make a purplish dye, and guano from bird droppings, which was used to make fertilizer.[53]

Although creative, Rolland's project was wildly impractical given the political environment, geography, and ongoing revolution. The rationale was clear and the project was perhaps theoretically possible, but it would have been a colossal and expensive undertaking. There were arid and boulder-filled hills. Carranza's government still possessed little real control over many of Mexico's states. It was in the middle of fighting civil wars and mediating between the belligerent forces battling in World War I. Although the Baja California peninsula, similar to Yucatán, had avoided the worst of the revolution's unrest, Chihuahua, for example, was still contested. The political leaders of the states along the eastern shores of the Gulf of California, mostly military men, were more concerned about holding on to power and exploiting the local populations than about long-term plans for the development of the wider region. Esteban Cantú, the military and political leader of northern Baja California, had no desire to collaborate with the Constitutionalists, whom he correctly saw as trying to destroy his rule. Most American business operators on the other side of the border would not support such an idea either; they backed Cantú, with whom they had lucrative business deals. Rolland's plan never left the drawing board. His essay did, however, bring renewed attention to securing Mexico's frontiers, developing hinterland resources, and countering the influence of foreigners and local, autonomously ruling military men.

Rolland's more general suggestions were more reasonable and gained traction among Constitutionalist leaders. In addition to stressing the canal, Rolland pointed to the need for increased colonization. He argued that the Mexican government should entice Mexicans who had immigrated to the U.S. state of California. Rolland's nationalism bordered on xenophobia. He lamented the large presence of Chinese and Japanese immigrants in northern Baja California. He argued that they were not assimilated Mexicans and thus were a threat to the Mexican nation. Rolland also reiterated the need to revise contracts with U.S. business operators and landowners and to increase communications links between the peninsula and mainland Mexico, something that the Carranza government acted on.⁵⁴

Problema de la Baja California also began a formal attack on Cantú that Rolland and other Carrancista intellectuals built upon for the next five years.⁵⁵ Cantú was a former Huerta supporter and an enemy of Carranza, only switching sides when Huerta's fall became imminent. Cantú had remained "neutral" during the subsequent civil war between the forces of Carranza and the breakaway armies of Emiliano Zapata and Pancho Villa. Most irksome to Rolland and others who supported greater national integration was Cantú's independence. He ruled part of the Mexican-U.S. border as if it were his own personal fiefdom, often in collaboration with powerful U.S. business operators. Rolland complained that Cantú wasted Mexican resources, implemented a byzantine tax structure, and worked mostly for his own personal gain, hindering the long-term development of the region.⁵⁶ Rolland would come down harder on Cantú in his subsequent writings. Cantú served himself, not the country, and he was ultimately a threat to Constitutionalist power and the integrity of the nation. Rolland may have been a Bajacaliforniano, but he had also become a fervent believer in the Mexican nation as a coherent whole. To him, Mexico had to become more unified or it would fall prey to powerful outside forces. This nationalist sentiment fueled Rolland's work in the peninsulas. Even as the Yucatecan experiment fell apart in the years following Alvarado's removal in 1918, Rolland's dream of creating a more just, prosperous, and unified nation-state continued to fuel his seemingly tireless ambition.

Rolland had used his skills diversely as a Carranza official. He had reported on educational trends in the United States, studied municipal governance, and written agrarian policy proposals. He worked as a top communications official during a critical part of Carranza's campaign against Pancho Villa and Emiliano Zapata. He had participated in a hugely important study on the petroleum industry. He had become a crucial member of Alvarado's state government, heading the agrarian reform program, assisting with Yucatán's feminist congresses, building port facilities, and designing infrastructure.

He had continued to build on past passions while learning new skills. Rolland retained his desire to improve the Baja California peninsula, developing enthusiastic if not always practical solutions to strengthen the region and its connection to Mexico. He made sure that national governments did not forget about Bajacalifornianos. Rolland took full advantage of his time in the United States to make new connections and acquire new abilities. His participation in U.S. progressive circles had benefited him and the Constitutionalist cause. While attacking the Church he had found faith in Henry George. He had become one of the most important Mexican publishers and propagandists in the United States, advancing the Constitutionalist cause and his own agenda. Rolland had become not only a talented engineer but also a capable writer with international connections. Combined, these attributes made Rolland valuable, adaptable, and potentially dangerous.

5

War and Peace

COMMOTION ECHOED OFF THE walls of New York City's Grand Central Station on the evening of July 24, 1916. Men and women of the American Union against Militarism hoisted a "flaming red and white banner" that read, "Friends of the United States and Mexico meet to bid good-bye to Señor Rolland, who is leaving to take part in the Joint Peace Commission at El Paso."[1] Carried on the shoulders of two of his Mexican compatriots, Modesto Rolland cut a path through the spectators to catch a 5:30 p.m. train to the Texas-Mexico border.

The United States and Mexico were on the brink of war. Pancho Villa had raided the border town of Columbus, New Mexico, the previous March, provoking the United States to invade northern Mexico and search for Villa in a campaign called the Punitive Expedition. The intervention had caused serious problems for the Constitutionalist leadership, who, like U.S. officials, wanted Villa dead but who once again could not politically afford to tolerate a U.S. invasion. Clashes between U.S. soldiers and Constitutionalist forces led some influential Americans to call for a full-scale war on Mexico.

As for the United States, the escalation of World War I in Europe caused great concern. Many people feared that the United States would soon be dragged into the conflict, and some U.S. officials feared that the belligerents, especially the Germans, could take advantage of the instability and growing anti-American sentiment in Mexico. The war in Europe had caught many U.S. intellectuals off guard. Germany, Britain, and France had all provided models of city planning and social work that progressives had praised. Many progressives blamed the bloodshed on capitalist greed. They hoped that despite the tragic consequences of the conflict they would be able to take advantage of the increased state planning of wartime to build community-based social models that would thrive after the war. This wartime socialism left a strong impression on Mexican progressive technocrats who were economically liberal but, in attempting to end the revolution, pushed for a strong government hand in economic planning and in improving social welfare.

At the same time, however, the brutality of the war in Europe also gave rise to doubts about the inherent goodness of nationalistic collectivism, efficient organization, and modern modes of technological production. Ambivalence crept in. Some groups in the United States, many of them led by women, including the American Union against Militarism, continued to promote progressive ideas but pushed energetically for U.S. neutrality and against violence.[2]

Rolland attempted to use his alliances with progressives to sell the U.S. public on the Constitutionalist cause and to thwart further war between the United States and Mexico. Rolland's skills went well beyond engineering. This versatility was not something unique to Rolland; a number of his Mexican engineering peers—the Urquidi brothers, Pastor Rouaix, Félix Palavicini, Alberto Pani—were persuasive writers, media operators, and bureaucrats. In addition to marketing his ideas for Mexico's territories, Rolland was serving as one of the leading voices for the Constitutionalists in the United States. In 1916 and 1917 he used his acumen to persuade U.S. progressives to campaign against a war with Mexico. In the process Rolland and his close associates swayed some of the United States' most prominent activists to back the Constitutionalists.

Rolland's endeavors highlight the Constitutionalists' shrewd foreign policy with respect to the United States. Carranza was a nationalist, but his government was not isolationist by any means; many of its members were savvy international actors, and they put serious effort into shaping public opinion and acquiring U.S. support. Technocratic diplomats were crucial to this endeavor. They spoke a language wrapped in "apoliticism" that many U.S. academics and members of the middle class welcomed. Rolland in particular was highly successful in obtaining the support of important individuals and organizations that influenced the policies of the Woodrow Wilson administration. Rolland demanded respect for Mexican sovereignty, but he openly incorporated foreign ideas, working to take advantage of transnational alliances to protect that sovereignty. Progressive ideas greatly affected his positions, and, in turn, he had a strong influence over his progressive counterparts in the United States.

U.S. and Mexican Forces Clash

The work carried out by Salvador Alvarado and Rolland in Yucatán had been possible only because of the relative peace in southeastern Mexico. Violence existed, but Yucatán was not nearly as war ravaged as parts of central and northern Mexico. While Rolland was redistributing land and building petroleum facilities on the southern peninsula, Gen. Álvaro Obregón and his army were clashing with Villa across northern Mexico, defeating him in important yet costly battles. In mid-1916 Rolland returned temporarily to the United States to promote the progress under way in Yucatán, but the violence in northern Mexico and a subsequent invasion by the United States shifted Rolland's efforts back to defending Constitutionalist actions and condemning U.S. intervention.

Earlier, in October 1915 in northern Chihuahua, a desperate Villa had pushed his fifteen hundred or so soldiers to the west and across the Sierra Madre Occidental. On the other side, in the state of Sonora, he had hoped to regroup and rebuild his forces. Villa encamped with his army outside the small border town of Agua Prieta, which was guarded by Constitutionalist forces under the command of Plutarco Elías Calles. It was there that Villa had obtained word that President Wilson had decided to back Carranza as the official head of the Mexican state. Furious, Villa moved on Agua Prieta. Little did he know that the U.S. government had allowed Constitutionalist forces to use an American railway along the border to move troops rapidly from Chihuahua to Agua Prieta. Villa sent waves of dusty soldiers against the town's defenses without success, wiping out much of his remaining troop strength. Villa and the remnants of his army fled once again.[3]

Hungry for revenge, Villa attacked the United States. He set his sights on Columbus, New Mexico. Approximately five hundred of his soldiers crossed the border in the dark hours of morning on March 9, 1916. Villa, acting on poor intelligence, believed that the U.S. Thirteenth Cavalry stationed in Columbus was not at full strength. The Villistas had caught the town off guard, but the number of soldiers was greater than reported, and they eventually repulsed the "bandits," killing approximately one hundred of them. Villa's men killed seventeen Columbus residents. According to an account by a U.S. Army officer, Lt. Jerome W. Howe, residents of Namiquipa, Chihuahua, later told American soldiers pursuing Villa that "Villa had made a speech there on his way to Columbus and said that he was going to strike a blow at the border, hurt the Americans' pride, and bring on intervention."[4] Americans remained proud, even belligerent, and if Villa wanted intervention, he certainly got it. Wilson sent a force of five thousand soldiers under the command of Gen. John "Blackjack" Pershing into northern Chihuahua to disperse and destroy Villa and his soldiers.

A little more than three months later, just before the dawn of June 21, 1916, members of the U.S. Tenth Cavalry, consisting mostly of African American "buffalo soldiers" under the command of a white officer named Charles T. Boyd, marched toward Carrizal, Chihuahua, ninety-two miles south of the U.S. border. Pershing's forces, after some initial success at dispersing Villa's forces, had not captured Villa himself, who had sustained injuries and gone into hiding. Boyd hoped at least to find some of Villa's lieutenants.

Villa, however, was not Pershing's only problem. The Constitutionalists grew weary of the American presence, and they resisted further encroachment into Mexican territory. Carranza and his generals warned Wilson and Pershing that they would not allow the incursion to expand. The U.S. intervention, Carranza argued, was destabilizing, and it fueled accusations that he could not defend Mexico from foreigners or, worse, that he was their collaborator.

Boyd found no signs of Villa's subordinates at Carrizal; instead he found the town guarded by Félix T. Gómez, a general in charge of a few hundred Constitutionalist soldiers. As the sun came up Gómez told Boyd that he would have to return to the north. Boyd, prideful and possessing a poor opinion of Mexican soldiers, was under the impression that if he moved on the town in battle formation the Mexican troops would scatter. Ignoring the advice of close advisors, Boyd ordered his men forward. When the U.S. soldiers came close, the Mexicans fired. A battle ensued. Boyd and nine other Americans were killed, dozens of men were injured, and twenty-four U.S. soldiers were captured. On the other side twenty-four Mexicans lay dead, including Gómez. Forty-three men had suffered wounds. Once news of the battle reached the United States, Wilson mobilized the U.S. National Guard and rushed troops to the border. Carranza organized volunteers in Mexico's northern cities for a possible "gringo invasion."[5]

A Problem-Filled Journey to Keep the Peace

William Randolph Hearst, a U.S. newspaper magnate with landholdings in northern Mexico and the inspiration for Orson Welles's film *Citizen Kane*, was one of the most outspoken proponents of a total invasion of Mexico. Articles in his *New York Journal* called for a war to "repress the Mexicans."[6] Hearst had long been influential. His newspapers had drummed up support for American intervention in Cuba in the 1890s. Hearst also had a personal interest in U.S. policy toward Mexico. He was upset about the Constitutionalists' seizure of his mother's ranch in Chihuahua, the Hacienda de Babícora, which stood at the heart of the conflicts between Carranza, Villa, and Pershing.[7] Hearst claimed that his motives were not personal, arguing that the U.S. government was failing in its "fundamental function" to "protect its citizens" and their properties. He chided "small Americans" for their calls for peace. His newspapers, Hearst wrote, were pushing the U.S. military to conquer Mexico not out of a desire for domination but to quell war and establish a genuine "advancement of peace and justice throughout the Western World."[8] Many of his readers bought into his "peace through war" narrative, but Hearst faced stiff resistance from antiwar activists and the Constitutionalists.

Political leaders in the Wilson administration and among the Constitutionalists were less eager for war. Wilson increased preparations for a possible invasion, but Hearst's demands provoked more perturbation than support from the president. Carranza meanwhile worked to calm tensions. The First Chief reached out to Wilson, stating that his administration did not seek war and was willing to negotiate but that he remained firm in his commitment to prevent the Punitive Expedition from expanding operations. As a sign of goodwill, he had the U.S. prisoners released.[9]

Rolland became involved with one of the U.S. groups most vocally opposed to war with Mexico: the American Union against Militarism. The organization's executive director, Crystal Eastman, was a prominent labor activist, peace proponent, and suffragist. She had become influential in progressive circles, whose members sometimes joined her own endeavors.[10] Lillian D. Wald served as president of the American Union against Militarism. Like fellow member Jane Addams, Wald was a prominent advocate of the "settlement house" movement.[11] The organization's leadership included a number of other prominent progressives as well, including Lincoln Steffens, Amos Pinchot, and John W. Slaughter.[12] It would soon include Rolland.

Rolland, along with his family, had just recently returned to New York. They had left Yucatán temporarily in order for Rolland to expand his campaigns to promote the single tax,

Alvarado, and Constitutionalist reconstruction more broadly. In early June a group consisting of Rolland, Carlo di Fornaro, and another Constitutionalist, Francisco Pendás, had started the Columbus Publishing Company.[13] Incorporated in New York with twenty-eight separate hundred-dollar shares, the enterprise was intended to improve the Constitutionalists' press operations. The corporation's stated goals included fostering better political and commercial relations between the United States and Mexico, as well as disseminating "the radical ideas of the present revolution in Mexico, endeavoring to establish justice for all classes of the inhabitants of Mexico, and freedom from tyranny and oppression of trusts, and to oppose private privileges."[14] Rolland, "representing the Constitutionalist government," bought 50 percent of the enterprise, and Pendás and di Fornaro obtained 25 percent each, further establishing Rolland's leadership in propaganda distribution in New York.[15] Under the corporation's umbrella Pendás would manage the monthly *El Gráfico* and Rolland and di Fornaro would run the Latin-American News Association.

The day following the Carrizal incident, the American Union against Militarism approached Rolland about participating in a conference between Mexican and American activists to help prevent further bloodshed. The group planned to hold the gathering on the border, in El Paso. Eastman must have found Rolland through progressive connections, possibly through Steffens, who had been at the Veracruz conference in December 1914 and who had been eager to work with Carrancista officials.[16] Rolland quickly jumped at the opportunity. When Rolland left New York City for El Paso on July 24, newspapers throughout the United States reported on the departure.[17]

The American Union against Militarism invited two other Mexican intellectuals to participate in the tentatively titled Peace Committee: Luis Manuel Rojas, the respected director of the Biblioteca Nacional, or National Library of Mexico, and Dr. Atl, who was an artist, labor activist, and editor of the newspapers *La Vanguardia* and *Acción Mundial*.[18] Rojas was intelligent, scholarly, and possessed strong organizational capacities. He would preside over the constitutional convention in the city of Querétaro the following year.[19] Dr. Atl, who knew Rolland, had created *Acción Mundial* as a response to U.S. intervention, but the newspaper, like *La Vanguardia*, also promoted itself as liberal, anticlerical, pan–Latin American, and "against all abuses."[20] European radicals' views had rubbed off on Dr. Atl, but he still held much in common with his childhood friends, some of whom, including the engineer Alberto J. Pani, became well-educated bourgeois advisors to Carranza.[21]

For the other side of the committee table, the American Union against Militarism invited David Starr Jordan, who had been president of Stanford University, Frank P. Walsh of Kansas City, and Secretary of State William Jennings Bryan. In addition to being the first president of Stanford, Jordan, the sole individual of the three who ultimately agreed to the invitation, was a prominent ichthyologist. He was a tall and "imperturbable" man, an anti-imperialist, and a peace activist.[22] In 1898 he had become vice president of the Anti-Imperialist League, and from 1909 to 1914 he was the chief director of the World Peace Foundation. Jordan had been quiet on the earlier U.S. invasion of Veracruz, but he became an important voice against those who called for war between the United States and Mexico after the battle at Carrizal, thus making him a useful ally for the Constitutionalists.[23]

Rolland and Jordan were the only members who made it to El Paso, which constituted a poor start for a committee tasked with stopping a potential war. The town was filled with angst. Some locals appreciated calls for mediation, but most people expected and even pushed for war. Residents called Jordan on the telephone, threatening him. Most of the warmongering public had no love for Rolland either. The mayor, Tom Lea, made it clear that he did not welcome peace activists. He told gathering crowds that he had ordered El Paso's police chief to tell Jordan to leave.[24] The mayors of Albuquerque and Santa Fe meanwhile telegrammed Jordan, telling him that the activists could meet in their cities. Jordan left for Albuquerque, though, according to him, he was not ordered out by the police.[25] Rolland left the following day.

El Paso may have seemed a fiasco for the American Union against Militarism, but it was not a complete failure. As Paul Kellogg, editor of *The Survey*, one of the leading social work journals in the United States, wrote the following month, "Frequently, in those days—with the national guard mobilizing and entraining for Texas camps, with Hearst news service fanning the flames and with the general press borne along by a wave of patriotic reporting—the only news item which got on the front pages of newspapers, in any way indicative of any effort to stem the tide of war, was the announcement from El Paso of Dr. Jordan's coming."[26] Rolland and Jordan's meeting had established a voice for peace.

In Albuquerque the two men hammered out a statement that was published in newspapers on June 28. It declared that there was "no rational cause for war," that the remedy for the violence in Mexico that had spilled over the U.S. border would not be further U.S. intervention, and that the U.S. government and public should support Carranza and his administration's work to regenerate Mexico. Exhibiting Rolland's influence, the statement cited the positive example of Yucatán and the adoption of the "New England 'township' system."[27] They also made it public that they had received word that Bryan and Walsh would not be able to join the committee. Instead they were to be replaced by Pinchot and Moorfield Storey, a pacifist and the inaugural president of the National Association for the Advancement of Colored People (NAACP), who would join them in Washington DC along with Dr. Atl and Rojas, who had only been delayed.[28]

The Peace Movement Finds Its Legs

Some of the affiliated members of the American Union against Militarism, including Irving Fisher, Gertrude Minturn, and Harry Overstreet, a professor at the City College of New York, had pleaded in person with President Wilson to not react hastily to the Carrizal incident and that mediation was the best approach. After listening to their resolution on arbitration, Wilson responded firmly that Carranza had to follow through on actions that proved that Americans and their property would "be safe from the depredations of Mexican bandits," that "acts must follow words."[29] Wilson did not want a full invasion of Mexico, but he also did not want to look soft in the face of possible war and the public anger over Columbus.

In the days before the reconvening of the Peace Committee on July 5, Jordan and Rojas collaborated on telegrams that they sent to Wilson and Carranza, informing them of the committee's desire for mediation between the two governments. They asked both leaders to put in place a ten-day armistice to let the atmosphere cool, "a breathing space" to allow time to establish a more formal joint commission. After a number of discussions and a letter from Carranza consenting to mediation, Wilson, too, agreed to attempt to find a solution diplomatically.[30]

The members of the Peace Committee meanwhile decided on a more formal name, the Mexican-American Peace Committee, electing Storey as chair and Rolland as vice chair. The committee put out press releases proclaiming that the calls for war were unjust and that the Wilson and Carranza administrations should work out a peaceful solution to the conflicts caused by the Punitive Expedition and border violence.[31] They addressed the need for dependable, honest sources of information about what was happening in Mexico, insinuating that this news would come from them. During subsequent meetings it also put forward a number of other lofty goals designed to build a mutually beneficial relationship between the peoples of Mexico and the United States. The committee, now including Fisher, who was a respected professor of political economy at Yale, and Leo S. Rowe, a scholar who studied Latin America and served as president of the American Academy of Political and Social Science, called for the creation of an exchange of teachers. It suggested further that American universities grant scholarships to Mexicans to study in the United States and to help specialists in Mexico create institutions to increase industrialization, improve agriculture, and develop competent teachers.[32]

Many of the Mexican-American Peace Committee's statements read as if they could have come from Rolland's Latin-American News Association. Despite their impressive educations and traveling experiences, the U.S. members remained relatively ignorant of Mexican affairs compared to their counterparts. The Americans relied heavily on the words of the Mexican representatives. And because Rolland was the only Mexican delegate who spoke English fluently, his influence had a huge impact on the committee's resolutions and statements.[33] The group reiterated the Constitutionalist reasoning behind the revolution and promoted the mutual agenda of U.S. progressives and Constitutionalist intellectuals: the need for more equitable taxes, the revaluation and redistribution of land, the restoration of free municipalities with elected councils, progressive labor reforms, and rights for women, including legal divorce. The committee also discussed the need to improve irrigation and for Mexico to gain control over its oil industry.[34]

While the Mexican-American Peace Committee continued its meetings, the official joint commission known as the Inter-American Peace Committee began to negotiate matters more formally. The committee's members included prominent politicians and scholars, including Secretary of the Interior Franklin K. Lane, attorney George Gray, and Rowe. Representing Carranza were some of his closest advisors: the famed lawyer Luis Cabrera and engineers Alberto Pani and Ignacio Bonillas. Carranza insisted that cooperation hinged on Wilson providing a specific timetable for the removal of Pershing's forces. The Americans argued that issues of border security and U.S. claims needed to be resolved first. The committee blamed Hearst and Republicans for warmongering, but the diplomats failed to come to any formal agreement. Further meetings took place, but the outcome did not change.[35]

The two peace committees were similar in that little of clear substance came from either of them. There was no immediate growth in educational exchange. At best they helped establish foundations for future partnerships. At the time the revolutionary unrest and fighting made establishing student exchange programs and U.S.-led development centers unfeasible. U.S. and Mexican representatives could not come to terms over withdrawing the Punitive Expedition, U.S. financial claims, or border security. Wilson, facing an election year, refused to defend Carranza, but he also refused to call for war. Wilson did downplay calls to violence, arguing that he was "not the servant of those who wish to enhance the value of their Mexican investments," a clear dig at Hearst.[36]

The peace committees did succeed in one important way—by their existence. They attracted widespread attention and the support of American peace activists and labor organizations countering calls for war. Rowe praised the Mexican-American Peace Committee, arguing that its influential but unofficial nature made it an excellent facilitator of a more balanced and beneficial understanding between Mexicans and Americans than the often vitriolic U.S. newspapers run by Hearst.[37] These committees' existence also provided something for Wilson to point to when asked about what he was doing to address the situation. Although the negotiations withered, they provided time for tempers to cool.

Perhaps what was most important for Carranza was that these committees, and the connections they deepened, helped the Constitutionalists achieve a long-standing goal: making their propaganda appear not to be propaganda. The committee and a number of newspapers reprinted Rolland's work, considering his pieces to be "dependable sources of information."[38] Invited by Rowe to speak in front of the American Academy of Political and Social Science, Cabrera, Pani, and Bonillas sold themselves as scientific and progressive humanitarians. Cabrera told the gathered crowd of academics that he had not come as a politician or a diplomat but as a member of the American Academy of Political and Social Science, which presented "a scientific interpretation of the facts which had been agitating Mexico."

Bonillas spoke in terms of reconstruction and evolutionary processes. Pani spoke of reestablishing order and improving sanitation through education and engineering, that there was a "necessary relation of direct proportion between the sum of civilization acquired by a country, and the degree of perfection attained by its sanitary organization."[39] Some of this was political theater, but most of it was genuine. Many Constitutionalist technocrats had been working on issues of education, infrastructure, and sanitation since before the revolution. U.S. progressives for their part saw themselves in these messages, and they increasingly worked with their counterparts serving the Carranza government. In this way U.S. progressives legitimized the Constitutionalist cause.

The Bullhorn of Anti-Intervention

Rolland penned his own anti-intervention pieces as well. He was constantly countering American yellow journalism, and he chastised the interventionist Hearst as misinformed at best and a self-interested liar and warmonger at worst. Claiming to be the first Mexican to refute Hearst, Rolland wrote, with a mixture of tact and literary flair, "You have, Mr. Hearst, cruelly attacked Mexico; you have roused great hatred and distributed so much venom, piling up such a mass of falsehood concerning our people, that even the most isolated Indian of our country is aware of the existence of a Mr. Hearst who owns many newspapers; knows that he is constantly maligned by that gentleman; and that the major part of that person's statements are false."[40] Rolland further contended that the Hearst family had not paid taxes on their Mexican properties and that the interventionist "false patriotism" of Hearst's newspapers was driven by a desire to maintain his family's wealth. In other words, Hearst was antidemocratic and wanted to block Mexico's social and political growth in order to protect his own interests.[41]

In an "Open Letter to the Honored President of the United States of America" Rolland continued to lambast American journalists and commercial interests for warmongering. He pleaded with Wilson not to cave in to belligerent demands. "You are well aware," Rolland wrote, "that when a weak people has to deal with that exploiting capitalism which knows no nationality, the latter seeks to employ as its agents the army and the navy, and condemns to death thousands of human beings deceived by false phrases of patriotism. The exploiters alone derive from the final catastrophe."[42] He argued that yellow journalists and a handful of U.S. capitalists in Mexico were willing to risk the deaths of thousands of people in order to siphon Mexico's wealth. Rolland continued, summoning his most colorful prose: "You know better than anyone else, Mr. President that the octopus of commercialism has captured us Mexicans, and that it tries to utilize the superior might of the American people in order to make futile our resistance, and afterwards more easily suck our blood."[43] Rolland supported beneficial exchanges, but he had grown to despise large international corporations and the tentacles they used to take advantage of developing nations.

In a piece titled "A Trial of Socialism in Mexico," published in the polemic journal *Forum*, Rolland wrote that he had recently been frequenting the American Club in Mexico City, becoming "fully convinced how charged the atmosphere was with the desire for intervention." Rolland warned readers that the United States would gain little through armed intervention but would instead "paralyze through perverted conceptions of humanity the socialistic regeneration that is progressing there."[44] Rolland hoped that he could persuade enough U.S. progressives who shared similar beliefs about social justice to create a counternarrative to that of the people who pushed for an invasion of Mexico.

While working as a bullhorn of anti-interventionism, Rolland and the Latin-American News Association also continued to promote Constitutionalist policies and actions. Other news outlets picked up their writings. In October 1916 the *New York Times* published a full-page article on Rolland, comparing him to the famous French writer Romain Rolland, to the Nobel Prize–nominated English writer H. G. Wells, and to the prominent American Unitarian minister and pacifist John Haynes Holmes. The article's author provided elaborate coverage not only of programs under way in Yucatán but also of the social and economic reforms in many of Mexico's states under Constitutionalist control.[45] Writers whose work appeared in other periodicals, including the *Herald of Gospel Liberty* and *Current Opinion*, praised Rolland for shedding new light on the "real" situation in Mexico, further legitimizing Rolland's statements while promoting his arguments in favor of land redistribution, municipal autonomy, and other progressive social reforms.[46] Rolland's press operations had become one of the most influential Constitutionalist outlets in the United States.

Critics

Rolland's work did not go without criticism. Writers for the *El Paso Herald*, which had earlier condemned the "unofficial peace conference" of Rolland and Jordan, argued that Rolland and his allies overstated the harm of American interests in Mexico, claiming that, to the contrary, U.S. capital in Mexico had "tremendously benefited Mexico."[47] In response to Rolland's *Forum* article, "A Trial of Socialism in Mexico," an American who gave his name only as "W.B." wrote that he had no qualms with Rolland's socialist ideals; instead W.B. argued that Rolland had overstated the Republican allegiances and interventionist sentiment of Americans.[48] Many of the Americans in Mexico were Texas Democrats. A sympathizer of Díaz and American petroleum interests, W.B. further argued that Rolland had exaggerated claims that Americans had gained immense wealth from exploiting Mexico's petroleum resources while providing little in return. Instead W.B. insisted that wealth extraction had ground to a halt during the revolution and that revolutionary factions had overturned the Díaz government's tax incentives. Although he sympathized with Rolland's "crusade against American intervention," the American argued that the Constitutionalist government could best avoid intervention by doing what Díaz did: preventing by military force the "outrages and raids on American citizens and properties."⁴⁹ There is no doubt that Rolland's essays were passionate and sometimes overgeneralized. There were prominent American capitalists who opposed U.S. intervention in Mexico: George Foster Peabody, John E. Milholland, and Daniel E. Burns, for example.⁵⁰ And there were indeed many Texas Democrats in Mexico, though Texas Democrats were a far different breed of Democrat than their East Coast progressive counterparts. Like many overgeneralizations, Rolland's claims were also not based solely on fiction. U.S. business operators were outspoken in their brazen calls for intervention, and as Rolland's U.S. critic made clear, so too were their sympathies for the prerevolutionary order.

Rolland's biggest critics were fellow Mexicans. The anarchist intellectual Ricardo Flores Magón, a zealous opponent of Carranza, applauded Rolland's calls for agrarian reform, economic freedoms, and less government deception. Flores Magón, however, was much more cynical than Rolland about the motives and capabilities of the Constitutionalists and found little difference between the old dictator Díaz and First Chief Carranza. And Flores Magón was right in some respects. Carranza was not going to end capitalist practices; he had no desire to do so. Far from ending government corruption, Constitutionalists became its foremost practitioners. Even Rolland admitted that many Constitutionalist military officers were corrupt. Flores Magón was wrong about another point, however. He argued that Carranza, Rolland, and their ilk would not be able to successfully reconstruct a Mexican government. The people, Flores Magón reasoned, would not settle for it now that they were in arms.⁵¹ Over the next three years the Constitutionalists would never fully succeed in quelling revolutionary violence, but they came close, reestablishing a functioning state while considerably expanding their control. Most Mexicans grew weary of the violence, and more people than Flores Magón, including radical laborers, admitted supporting the Constitutionalist government.

The worst of Rolland's enemies were even closer to him. Rolland's blunt personality and determined views did not sit well with all of his colleagues. He was a harsh boss and difficult colleague.[52] Pendás in particular grew to despise Rolland. In January 1917 Pendás lobbied Carranza to remove Rolland as head of the Latin-American News Association and as the leading shareholder of the Columbus Publishing Company. Pendás argued that Rolland was profiting from the press and conducting a hidden campaign against the First Chief. Pendás sent photographs of letters that Rolland had penned to Alvarado congratulating the latter's advancement of land redistribution around the Maya town of Uxmal, which was against Carranza's wishes. Comparing their undertaking to the French Revolution, Rolland had pushed Alvarado to take his opportunity as governor to be bold and to fight off the "spirit of conservative reaction" within the ranks of the Constitutionalists. According to Rolland, that reactionary spirit was already "on the horizon."[53]

Pendás made strong accusations against Rolland's personal character as well. Most disturbing was Pendás's accusation that Rolland was "savagely" abusive. According to Pendás, Rolland "brutally beat his wife"—who was pregnant with their fifth child, Carmelita—and mistreated employees of the Latin-American News Association. Pendás claimed that he and other staff members had witnessed Garza de Rolland and Rolland get into an argument after the former stormed into the office, upset with Rolland's absence and behavior. According to Pendás, a secretary had once called the police and it was his actions alone that stopped them from arresting Rolland, thus staving off a public relations disaster. Of course, his supposed actions also made certain that there was no record to verify that the misdeed ever happened.[54]

Rolland carried on extramarital affairs and was at the very least verbally abusive, but it should be noted that Pendás had a serious ax to grind. As the head of the Columbus Publishing Company, Rolland presided over Pendás's operations. Pendás disagreed with Rolland's philosophies and pleaded for Carranza to remove "Rolland's dictatorship." Pendás possibly exaggerated Rolland's personal faults to make up for a weak case against

Rolland's political loyalties. It is true that Alvarado and Rolland had their eyes on their current alliance and political futures. Rolland worked as Alvarado's commercial agent in New York, setting up agreements for the purchase of henequen.[55] Rolland promoted his own Georgist policies, yet he had no intention of bringing about the First Chief's demise. There is also no other evidence that Rolland was physically abusive, that he "savagely" beat Virginia. His grandchildren recalled no stories of him being physically abusive or seeing any abuse themselves. The truth of the matter, like so much that has passed into the dustbins of history, is unclear.

Rolland was aware of Pendás's distaste for him and of Pendás's accusations. In response Rolland wrote regularly to Carranza, defending his actions and emphasizing his importance. Rolland reiterated that under his leadership the Latin-American News Association remained critical for resisting calls for U.S. intervention, justifying the revolution, and promoting the fruits of Constitutionalist reconstruction.[56] Rolland was also quick to point out his operation's mailing list of twenty-five thousand prominent people and organizations.[57] Under his supervision, Rolland reminded Carranza, the Latin-American News Association was efficient, influential, and worthwhile. He accused Pendás of being the greedy partner and argued that Pendás was slanderous and corrupt. Rolland admitted that he had used money from the Constitutionalist government, as well as personal funds from Alvarado, in order to keep the operation going, but he maintained that the money was spent honestly.[58]

In regard to *El Gráfico*, Rolland worked to undermine Pendás's directorship, asking Carranza to cut Pendás's support. After stating that Pendás had acquired thousands of dollars from an unnamed woman close to Carranza, Rolland persuaded di Fornaro to sell his shares in the Columbus Publishing Company to him. Rolland then controlled 75 percent of the stock, providing him more control over the company and *El Gráfico*. Rolland shared all of this information with Carranza. But Rolland did not want the periodical itself shut down. He thought it could be immensely important, and he begged Carranza to send $2,000 in order to fund the operation through April 1917. Rolland stressed that he was the man with the "firmest convictions and loyalty to the [Carranza's] movement" and that he wanted to turn the newspaper into a "true voice for all of Latin America" to counter savvy "filibusters."[59] Circumstantial evidence suggests that the First Chief ultimately sided with Rolland or that Rolland at least won out in the conflict. In his executive briefings Carranza continued to receive updates on Rolland's work, which showed Rolland as capable and influential in U.S. circles.[60] Rolland continued to head the Columbus Publishing Company until at least 1919.[61]

Publication and leadership changes at *El Gráfico* are also telling. Pendás had relinquished his management of *El Gráfico* by the end of February 1917. Rolland subsequently published articles in the April 1917 and January 1918 issues. He had never written directly for the periodical when it was under Pendás's directorship. Rolland wrote an article praising Carranza, but the journal also gave more attention to Alvarado and development in the Yucatán, including Rolland's work on petroleum facilities and his drive to establish a single tax.[62] This increased attention on Alvarado shows an attempt to paint him as a potential candidate for the presidency in 1920, when Carranza was slated to hand over the reins of executive power. But if Carranza had issues with Rolland, especially in 1916 and early 1917, those issues were not so consequential as to outweigh Rolland's advancement of the Constitutionalist cause. Rolland remained a prominent figure in promoting the Constitutionalist agenda, maintaining Carranza's support despite their disagreements.

Like all people, Rolland had faults and virtues. He was aggressive. He was stubborn. He often treated his wife poorly and spent little time with his children. But he possessed organizational talent and an untiring work ethic. He was a practical planner but also a man of imagination and ideals. He genuinely believed that his family would forgive his absences and temper once the fruits of his endeavors were harvested in a Mexico that even the United States would desire to emulate. And as his work with the Latin-American News Association and the American Union against Militarism clearly exemplifies, his influence transcended his engineering skills. For better and worse, he meshed his planning and building talents with rhetorical persuasion. He was a renaissance man, perhaps more so than many of the engineers who would follow him with their more specialized focuses in the latter half of the twentieth century and the early twenty-first century.

The examination of Rolland's life in the United States in 1916 and 1917 shows something more than revolutionary engineers becoming enmeshed in politics and prose. Rolland was a prominent figure in preventing the expansion of war between the United States and Mexico. His actions may not have been the determining factor in influencing the Wilson administration's decision not to expand operations in Mexico, but they did not go unnoticed. Wilson knew about them. Rolland received significant attention from the press, allowing him to influence U.S. public opinion and to counter calls for war. Rolland's partnerships are also a clear display of the strong relationship between Constitutionalist intellectuals and U.S. progressives. In a sense Rolland and his colleagues used them. Rolland, through personal negotiations, pamphlet production, and press releases, galvanized support from many of America's progressive leaders, who in turn helped make him an "objective" voice. Rolland sold the Constitutionalists as the equivalent of middle-class progressives, as the Mexicans most like U.S. intellectuals, or even President Wilson for that matter. Many progressives in the United States in turn saw Rolland and his colleagues as their counterparts. In their magazines and speeches they portrayed the Constitutionalists as revolutionaries who were refined and educated. To paraphrase Mary Hunter Austin, a famous author and environmental activist of the time who had an article published by the Latin-American News Association, the Constitutionalists would continue the "civilization" of Mexico that Díaz had started, but with a greater emphasis on education and justice.[63] In other words the Constitutionalists were for many U.S. progressives the representatives of the middle class and the middle ground between the dictatorial but civilized Díaz and the virtuous but ignorant Indian peasants. And as unsophisticated as overgeneralizations typically are, this image solidified significant support from prominent Americans, greatly aiding the Constitutionalist cause. With the Constitutionalists more firmly entrenched as the political leaders of Mexico by 1917, Rolland continued to promote his vision of reconstruction. He also braced himself for the storms of political change hovering on the horizon.

FIG. 1. Students at the Colegio Rosales in 1898. Modesto Rolland is on the far left of the fourth row. Photo courtesy of Jorge M. Rolland C.

FIG. 2. Mexico City, ca. 1890. The occasional car would have made its way onto the streets by the time Rolland arrived in 1901. Wikimedia Commons. Scanned from *A Photographic Trip around the World* (Chicago: John W. Illiff & Co., 1892).

FIG. 3. Rolland as a young engineering student, 1905. Photo courtesy of Jorge M. Rolland C.

FIG. 4. The de la Garza family, 1908. Rolland's soon-to-be spouse, Virginia de la Garza, is standing on the far right. Photo courtesy of Deanna Catherine Wicks.

FIG. 5. Portrait of de la Garza during her engagement to Rolland, 1907. Photo courtesy of Jorge M. Rolland C.

FIG. 6. Workers building the Xochimilco–Mexico City aqueduct, 1908. Photo courtesy of the Archivo Histórico del Palacio de Minería.

MODESTO C. ROLLAND.
INCENIERO CIVIL
(E. N. de I.)
ESPECIALIDAD EN CONCRETO ARMADO.

22 metros de claro.
Salón de Exposición de los automóviles F. I. A. T. Calzada de la Reforma.

Avenida de la Independencia No. 8,   MEXICO, D F.

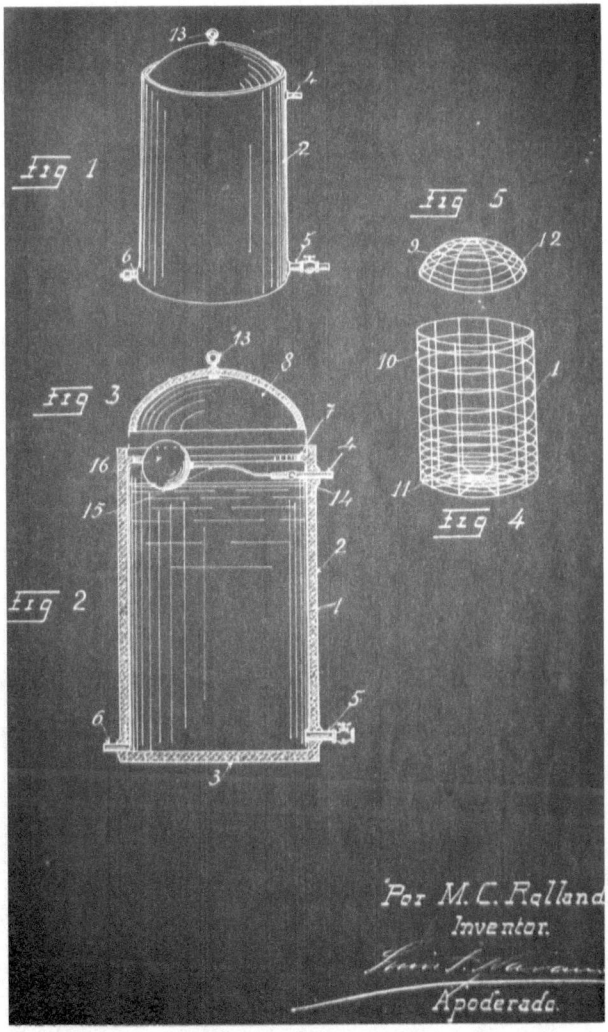

FIG. 7. (*opposite top*) New members of Porfirio Díaz's last cabinet, including Rolland's mentor, Manuel Marroquín y Rivera. Library of Congress, Prints and Photographs Division, George Grantham Bain Collection, LC-DIG-ggbain-09396.

FIG. 8. (*opposite bottom*) Ad for Rolland's concrete workshop, ca. 1911. Photo courtesy of the Archivo Histórico del Palacio de Minería.

FIG. 9. (*above*) A blueprint for Rolland's patent filing for a reinforced-concrete water tank, 1913. Photo courtesy of the Archivo General de México.

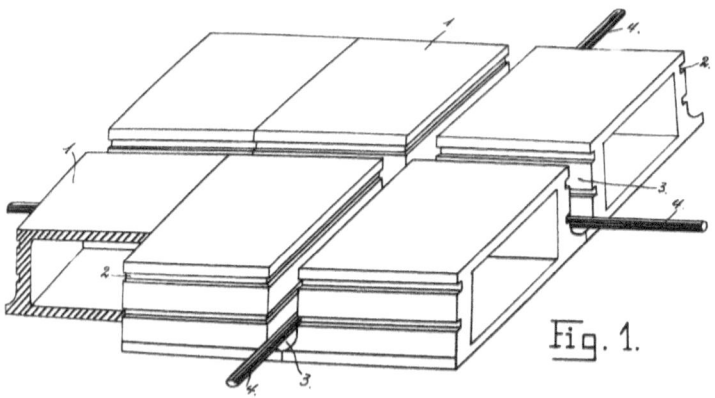

FIG. 10. A diagram of a reinforced-concrete process Rolland patented, 1913. Photo courtesy of the Archivo General de México.

FIG. 11. The Rolland family, 1913. Photo courtesy of Jorge M. Rolland C.

FIG. 12. Street scene during the Ten Tragic Days in Mexico City, 1913. Library of Congress, Prints and Photographs Division, LC-USZ62–107789.

FIG. 13. Military cadets at Chapultepec Castle, 1913. Library of Congress, Prints and Photographs Division, Harris & Ewing Collection, LC-DIG-hec-02056.

FIG. 14. U.S. soldiers raising the American flag during their occupation of the city of Veracruz, 1914. Library of Congress, Prints and Photographs Division, George Grantham Bain Collection, LC-DIG-ggbain-15834.

FIG. 15. Venustiano Carranza's cabinet members and close advisors, 1915. From left to right: Juan Sánchez Ascona, José N. Macias, Luis Manuel Rojas, Pastor Rouaix, Luis Cabrera, Venustiano Carranza, Ignacio Bonillas, Mario Méndez, Juan Neftalí Amador, Modesto Rolland. Photo courtesy of the Centro de Estudios de Historia de México.

FIG. 16. Rolland with Gen. Álvaro Obregón shortly after Obregón lost his arm in battle, ca. 1915. Photo courtesy of the Centro de Estudios de Historia de México.

FIG. 17. Henry George, ca. 1880. George's ideas on taxation and land use had a profound influence on Rolland. Library of Congress, Prints and Photographs Division, George Grantham Bain Collection, LC-DIG-ggbain-37132.

FIG. 18. U.S. progressives Crystal Eastman and Amos Pinchot, ca. 1915. Library of Congress, Prints and Photographs Division, George Grantham Bain Collection, LC-DIG-ggbain-21959.

FIG. 19. U.S. progressive Lincoln Steffens, 1914. Library of Congress, Prints and Photographs Division, George Grantham Bain Collection, LC-DIG-ggbain-15929.

FIG. 20. Illustration representing U.S. frustration over the Punitive Expedition, with an American soldier (probably representing Gen. John Pershing) tied (figuratively by U.S. policy and by Venustiano Carranza) to a cactus, while an injured but dangerous Pancho Villa sneers and Carranza and a donkey look on from their perch on a distant volcano. William Allen Rodgers, *New York Herald*, November 26, 1916, 2. Library of Congress, Prints and Photographs Division, Cabinet of American illustration, LC-USZ62-130777.

FIG. 21. Illustration by Nelson Harding commenting on Carranza's use of propaganda to influence American policy, published in the *Brooklyn Eagle* and *Current Opinion*, 1916.

FIG. 22. Adolfo de la Huerta, interim president of Mexico, 1920. Wikimedia Commons. National Photo Company Collection.

FIG. 23. Obregón around the time he assumed the presidency, ca. 1920. Library of Congress, Prints and Photographs Division, George Grantham Bain Collection, LC-DIG-ggbain-25501.

FIG. 24. The El Buen Tono booth at the 1923 Mexico City Grand Radio Fair, with Modesto C. Rolland standing at the far left. Photo courtesy of Jorge M. Rolland C.

FIG. 25. Ad for the free ports in the Mexico City newspaper *El Demócrata*, 1924.

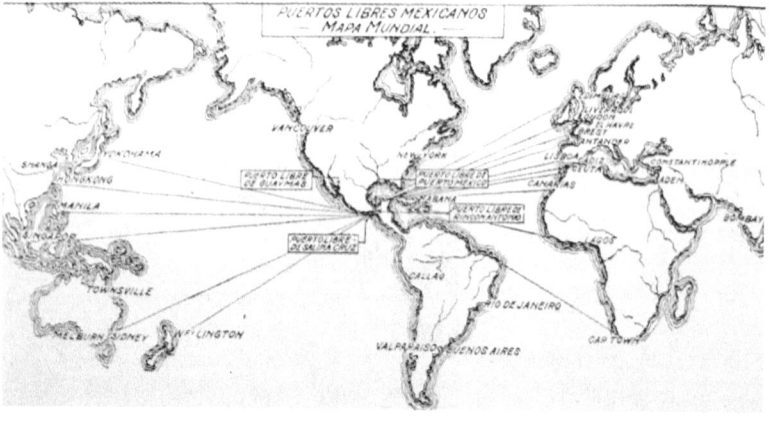

6

Transitions

ON DECEMBER 1, 1916, Venustiano Carranza delivered a speech to officials who had gathered in the city of Querétaro to write a new constitution for Mexico. Before launching into a lesson about the Constitution of 1857 and how politicians had abused it—which he did with some regularity—Carranza praised his constituents for fulfilling a promise he had made to bring together the Constitutionalists' decreed reforms into a new and binding constitution for the Mexican people. Carranza called for a government with checks and balances. He called for guarantees for free municipalities. Just as important, however, is what Carranza did not say. He said nothing about agrarian reform, workers' rights, or the nationalization of subsoil resources. The vision he laid out was simple and relatively conservative.[1]

Carranza allowed only people at least nominally loyal to his cause to participate in the creation of the new constitution. Most of them were more or less middle class, though the majority of them held no professional title or degree. Twenty-one of them, nearly 10 percent, were engineers, including some of the most influential architects of the resulting document. There were other professionals, too, including lawyers and doctors. Most of the participants were young, below the age of forty.[2]

Exemplifying the divisions within the Constitutionalist ranks, the end product of the convention was far more radical than Carranza had desired. It was among the most radical constitutions ever produced anywhere in the world. The Constitution of 1917 provided the state greater legal power in the realms of social and economic reform. Most celebrated (and chastised by foreign business leaders) were Articles 27 and 123. The former made the state the sole owner of Mexico's subsoil resources and gave the state the right to nationalize lands it considered necessary for the national good. The article restricted foreign ownership of properties on Mexico's coasts and put in place the legal underpinning for land redistribution. Article 123 took up progressive labor issues. Among other things, it established a minimum wage, an eight-hour work day, the end of debt peonage, the right to strike, and limitations on child labor. The engineer Pastor Rouaix was a key figure in the writing of both articles.[3]

Officials in the Southern District of Baja California had selected Modesto Rolland as one of their representatives for the constitutional congress in Querétaro, but others served in his place. The reasons he turned down the honor remain unclear; he stayed in New York City churning out publications for the Latin-American News Association, including Carranza's opening speech and essays discussing his policy ideas. Rolland did aspire to influence his peers at Querétaro. Rolland asked officials in Yucatán and Baja California to send information on educational progress to the constitutional assembly. When he returned to Mexico, Rolland stayed mostly in Yucatán, continuing his work with Alvarado.[4]

Despite having made significant strides in forming a viable state, Carranza's forces had still not secured all of Mexico—Pancho Villa, Emiliano Zapata, and Félix Díaz all retained small armies opposed to the Constitutionalists—but the constitution and Carranza's acquisition of the title of president went a long way toward consolidating their rule. The Constitutionalists possessed Mexico City, the most important ports, and many of Mexico's most profitable regions. They controlled most of the communications networks. They had acquired formal recognition from the world's most powerful governments, including that of the United States.

The consolidation of Constitutionalist control had a significant impact on Rolland's endeavors. With greater U.S. support there was less need for Rolland's media services in the United

States, though he continued on, in an increasingly limited capacity, until 1919. Carranza removed Salvador Alvarado from his position in Yucatán in early 1918, pressing him back into military service elsewhere.[5] Rolland worked with Alvarado's successor, Carlos Castro Morales, but without the strong backing of Alvarado, Rolland's ability to influence agrarian policy waned. Instead he continued to work on less controversial infrastructure projects.

Recognizing his limited capacity to effect change abroad and in Yucatán, Rolland returned to Mexico City in 1919. There he managed an Alvarado-financed newspaper, participated in Mexico City engineering organizations, and joined a government commission tasked with reporting on conditions in northern Baja California. Political tensions increased again as 1920 drew near. Constitutionalist leaders selected that year for the first peaceful presidential transition since the start of the revolution. Political transitions had not gone well in Mexico since Díaz's fall from power. In addition to addressing the inherent difficulties of improving material conditions, Rolland had to find ways to navigate hazardous political waters and to keep himself relevant in top political circles while not appearing overtly political.

Rolland's career, however, is chock full of examples of how engineering projects and politics are intertwined. This period of Rolland's life is no different. It provides an example of just how much the work of engineers was influenced by Mexico's dramatically shifting political conditions. It also shows how gifted Rolland was at navigating these political obstacles and how luck was on his side. Engineers like Rolland, who managed to remain influential during extreme political turmoil, had to be astute political actors. The survival of their careers and dreams often depended on remaining important to top political leaders. Likewise, since Mexico possessed a limited number of highly capable engineers, politicians needed these technical specialists to help realize their own political agendas. Yet these experts and politicians regularly frustrated each other. It was a complicated and dangerous dance. And when added to the immense and complicated compromises that military leaders, industrialists, peasants, urban laborers, foreign interests, regional bosses, and Mexico City elites had to make with each other to bring about a semblance of peace, this dance made completing major projects difficult and long-term planning almost impossible.

Last Writings from New York

Although Rolland did not participate directly in the constitutional congress in Querétaro, he did attempt to influence its members through the Latin-American News Association and essays published in Mexican newspapers. In a work he titled *Carta a mis conciudadanos* (Letter to my fellow citizens) Rolland reemphasized the need for an improved tax system, agrarian reform, and a government-directed petroleum industry. He also argued that Mexicans would have to learn from Americans and Europeans in order to save themselves from these same powerful industrial societies. Rolland's concerns about the future of the revolution boiled over in the essay. He did not attack Carranza by name, but Rolland was obviously frustrated by the lack of rapid and wide-ranging improvement. At the very least, Rolland thought, the Carranza administration could be bolder in its initiatives. It was not doing enough to develop Mexico, to protect its territories, and to bring about social justice on a national scale. He pushed for the authors of the new constitution to do more.[6]

Carta a mis conciudadanos was part of a shift. Although Rolland continued to operate the Latin-American News Association and remained the leading stockholder in the Columbus Publishing Company, he put more effort into influencing policy in Mexico City during the years of Carranza's official presidency (1917–20). Rolland also put more time into assuring his own professional survival.

Rolland's advice and grievances were not, however, limited to Mexico. He had grown frustrated with Americans, too. Following Frank Pendás's departure from *El Gráfico*, Rolland printed in the journal an article on Mexico's neutrality during World War I. Suggesting that Americans were exploitative and insensitive,

Rolland fumed, "The American people are so entirely ignorant of the habits, temperament, and conditions of Latin-American countries." The war, Rolland continued, was "nothing more than the impact of great Anglo-Saxon and German trusts." He wrote that Mexico and other Latin American nations would not participate in a war in which the belligerents were the very nations that had exploited their homelands. Rolland started the essay as a criticism of U.S. ignorance, but he turned it into a critique of the Western social order in general: "We observe that religious beliefs as well as material civilization have become a complete failure, and that humankind is compelled to destroy itself in the most frightful manner under the urge of powerful commercial interests."[7] Rolland was not alone in his frustration with Europe and the United States. His message would have resonated with many people in Latin America. World War I had burst open the ugly side of imperial capitalism and the technological tools that developed with it. People in regions outside of Europe began to doubt more regularly Europeans' own claim to superiority.

Yet Rolland still believed that humans were ultimately on a path of betterment. It was just that the old imperial orders were losing their luster. It would be in places like Mexico that real social progress would become a reality, he believed. It should be further noted that Rolland's criticism of Western materialism was more a critique of greed than of development. Almost every professional endeavor Rolland had worked on, and would continue to work on, was to improve Mexico's infrastructure and material conditions.

World War I may have left Rolland somewhat jaded in his views on the United States and Europe, but it didn't ultimately lessen their influence on him. He wanted a Mexico happier and less greedy than the United States, but he realized that the Colossus of the North would not become any less colossal in the near future. Rolland wrote, "We [Mexicans] must assimilate what is necessary to establish a national evolution in every way. If we cross our arms we will have no choice but to be cast out of our land and our homes and go to the field to sing our sad ballads as the last manifestation of our national spirit."[8] Mexicans would have to build their country in the face of massive American influence and material wealth. Rolland proposed that the Mexican government send thousands of young Mexicans to study U.S. and European systems of production so they could return to "conquer Mexico" for Mexico.[9] Of course in Rolland's mind these nationalist, transnational students would gain the skills required to strengthen Mexico, but they would build up their country without the greed displayed by the American titans of enterprise. These students would be used to build a capitalist Mexico, but they would create a socially conscious form of capitalism, which Rolland and many other progressives around the world referred to as socialism. In other words, Rolland promoted a state-influenced form of capitalism based on the welfare of society. And much like the earlier Cuban nationalist and visionary José Martí, Rolland would throughout his life wrestle with alternating distaste and admiration for the United States.

Waning Days in Yucatán

Well into 1918 Rolland continued to work off and on in Yucatán while publishing articles on agrarian reform, the single tax, autonomous municipal governance, improved education, petroleum development, increased infrastructure, and the "problem of Baja California." Living up to his liberal background and Henry George land-tax creed, Rolland collected, on behalf of the Yucatecan government, rent from companies using lands expropriated from the Church.[10] He also spent considerable time exploring for oil.[11] As mentioned previously, he did not find any, though future petroleum developers building on his initial investigations would be more successful. During these explorations Rolland argued that petroleum in Yucatán, once found, produced, and refined, should be used to advance irrigation efforts, not only in Yucatán but in the arid northwest as well. On the Baja California peninsula the land was dry, but Rolland argued that there was water underground that could be pumped to the surface using gas engines, transforming his home region into a wealthy and productive territory like the U.S. state of California to its north. This would help entice Mexican migrants in the United States to return and colonize Baja California, lessening the influence of foreigners while increasing trade and prosperity. The petroleum would get to Baja California, Rolland argued, via train. Railway construction had increased in Yucatán under Alvarado, and Rolland wanted to connect these lines to the Tehuantepec Railway. This railway connected the Gulf of Mexico to the Pacific, crossing Mexico at the nation's narrowest point, the Isthmus of Tehuantepec, allowing for the transport of oil from the oil fields around Veracruz and, he hoped, Yucatán. By doing this, Mexicans would develop the nation's hinterlands and increase trade in the Pacific, lessening the country's dependence on the United States and Europe.[12]

It is unclear if Rolland, Alvarado, Luis Cabrera, the engineer León Salinas, or someone else first promoted the government taking over the Tehuantepec Railway and its connecting ports. The move appears to have been partially influenced by the Development Company of the Southeast. By 1916 Alvarado had been promoting the connection of Progreso and Yucatecan towns to Tehuantepec via railroads. Cabrera, then secretary of finance in the Carranza administration, promoted Alvarado's plans for railway expansion and Rolland's work building petroleum facilities in Progreso.[13] Alvarado contended that the increased infrastructure would help build local and regional economies, as well as expand the production and markets for henequen, chicle (used for gum), timber, coffee beans, cacao, flowers, and many kinds of fruit. He ambitiously wanted to transform the ports of Puerto México (later Coatzacoalcos), Salina Cruz, and Progreso into first-rate centers of trade. He argued that these developments would allow for the growth of a national petroleum industry and, he hoped, industrial manufacturing enterprises, all while better unifying the Mexican nation.[14] Of course Rolland had been arguing for the same things. In fact he, unlike Alvarado, was actually overseeing port improvements and constructing petroleum facilities. Carranza, however, put an end to

Alvarado's aspirations and along with them many of Rolland's plans for Yucatán. Alvarado had prepared to run for election as civil governor, but Carranza ordered the general to take on military tasks elsewhere in southern Mexico.

Rolland nonetheless made some noteworthy contributions to Yucatecan society during Alvarado's last months as governor and even after his ouster. Rolland finished the Petroleum Station Terminal and work on the pier in Progreso, one of the only such facilities in existence in Mexico.[15] Building on his previous experience as a communications official, Rolland, while manager of the Petroleum Department of the Development Company of the Southeast, aided in the expansion of the state's telegraph and radiotelegraph network. The Alvarado administration had constructed the territory's first radiotelegraph station in 1916. Alvarado praised the "limitless utility of this modern invention."[16] During the government of Castro Morales, Rolland bought receptors for the Petroleum Department. He also advised the new governor on the importation and use of equipment and aided in the development of a telegraphy school.[17]

Rolland also brought new construction technologies and methods to Yucatán. He started a business building one- and two-story homes out of reinforced concrete. He hoped to improve housing across the peninsula. He established a small factory to produce concrete blocks and ultimately sold four different models of small, "affordable houses." The houses were well built and simple but with windows and nice ornamentation. Rolland does not appear to have sold many of them, but his work helped popularize modern concrete-block home construction in the state.[18]

Rolland's changing roles in Yucatán demonstrate that when he ran into insurmountable obstacles, including Carranza's rejection of his agrarian policies, Rolland moved on to other projects where he could make genuine progress with less resistance. That pattern also shows his inability to follow through with his long-term plans for development, most often because of these political roadblocks.

Rolland's schemes for petroleum and infrastructural development ran into their own massive obstacles. First, Rolland needed to find petroleum, something he never did, at least on a substantial scale. Second, there were still parts of Mexico, including Tehuantepec and Baja California, where Carranza's rule was challenged. As eager as the First Chief was for political consolidation and as Rolland was for advancing Mexico's material welfare, the reality on the ground kept their goals at a distance. And although Rolland does not appear to have met serious political resistance to his plans for developing petroleum production, storage, and transport in Yucatán, unforeseen problems awaited his plans for irrigation, the Tehuantepec Railway, and development in Baja California. Each project created new political gauntlets.

The Return to Mexico City

Rolland returned to Mexico City in 1919 to help establish and direct a new Alvarado-owned newspaper: *El Heraldo de México*. Rolland described it as a means of "preaching the new ideas on the single tax and the modern forms of municipal administration with the Referendum, the Initiative, and the Recall."[19] It would serve as the capital's progressive newspaper. Alvarado also financed it to promote himself. He was considering running in the upcoming presidential election.

As for Rolland, the newspaper allowed him to become a prominent voice in Mexico City circles again. His ideas had been making their way in from New York and Yucatán, but now he could present them in the capital immediately and in greater detail. *El Heraldo de México* published pieces about a number of ideas near and dear to Rolland: land and oil policies, Samuel Gompers and transnational labor cooperation, the sins of Esteban Cantú in Baja California, and the achievements of Alvarado.[20] The newspaper's pieces on Baja California and the "petroleum problem" are identical to Rolland's arguments in his own essays.[21] The newspaper articles on oil refer to petroleum development and ownership as the gravest problem in Mexico's history, affecting the country's honor, spiritual autonomy, inalienable rights, and national existence. *El Heraldo de México* pushed for a strong implementation of Article 27 of the Constitution of 1917. Foreign oil companies, Rolland argued, had gained far too much influence in Mexico, draining Mexican wealth while putting the country at risk for future intervention.[22] As for Baja California, Rolland had not eased up on Cantú or the need to develop the western peninsula. Until the region was better incorporated it would remain threatened by usurpation by the United States, and its underdevelopment would burden not only Bajacalifornianos but all Mexicans.

Living in Mexico City, Rolland reimmersed himself more deeply in the capital's politics, engineering circles, and elite society. Journalists and editors from the various news agencies in the city put on "lunch-champagnes" for Rolland, Alvarado, and the other contributors to *El Heraldo de México*.[23] He attended a ball honoring the French community.[24] At other events he rubbed shoulders with prominent officials, including the governor of the Federal District and the undersecretary of industry and commerce.[25] Rolland built closer ties to Francisco J. Múgica and Heriberto Jara, acquaintances and important military commanders and politicians, both of whom were themselves visionaries when it came to Mexico's future.[26] These contacts would prove important later in ways Rolland did not foresee.

Rolland became a leading figure in the recently established Centro de Ingenieros, or Engineers' Center. The organization's membership included people from all over Mexico, but it was closely connected with the National School of Engineering. At the center many of the country's top engineers bounced ideas off each other, collaborated, and mixed with politicians. In an official statement issued by the organization, its engineers were "serving the government so throughout the country engineering projects could increase wealth, improve sanitary conditions, and beautify the cities."[27] In August the center hosted a ball, which Rolland attended along with prominent politicians, doctors, and his engineering colleagues and their spouses or girlfriends. He was one of the few attendees noted in the press as not having someone with him.[28] He spent Christmas attending another Engineers' Center celebration.[29] The Engineers' Center became a home of sorts for Rolland. Most of his undertakings had some link to the group—it provided recruits for his causes, the organization promoted his ideas, and he used the group's space for technological displays and for discussions on the intersection of policies and engineering projects.

The Report on the Northern District of Baja California

Rolland, however, rarely stood still. His critiques of Esteban Cantú in *El Heraldo de México* drew not only from his prior writings but also from travels to Mexicali, Tijuana, and other parts of northern Baja California. Aware of Rolland's interest in the region and his skill as an observant planner, the secretary of finance sent him and fellow Constitutionalist officials Rafael N. Millán y Alva, Fernando de Fuentes, and Miguel López to the Northern District of Baja California to report on its economic and political conditions.[30] It was a turbulent moment. Not only was Cantú resisting the authority of the central state, but the run-up to the 1920 presidential transition was also well under way and there looked to be a real possibility of renewed civil war. The expedition to the Northern District was called for because of the Organization of the Federal District and Territories Law (1917), and it followed up on a previous commission visit to the Northern District of Baja California headed by Rolland's colleague, the prominent engineer Pastor Rouaix.[31] The goal of all of these efforts was to bring Baja California under Constitutionalist control and to undermine Cantú's grip on the region. Although Millán y Alva was the head of this 1919 inspection, it was Rolland's conclusions, published individually in multiple articles in *El Heraldo de México* and together as *Informe sobre el Distrito Norte de la Baja California* (Report on the Northern District of Baja California) that most dramatically influenced political leaders and latter-day scholars.[32]

Rolland was well suited to this task. Unlike Millán y Alva, Rolland was from Baja California. Rolland wrote passionately about the region, and his distaste for Cantú had been vociferously expressed since 1916. Rolland had also played a prominent role in reorganizing the economy of Yucatán, which was similar to Baja California in that it was a hinterland peninsula that possessed wealth. As with Yucatán, the Carranza government stood to gain substantial economic benefits from better incorporating Baja California. Whereas Yucatán possessed the wealth-producing henequen industry, Cantú's government had made handsome gains from cotton and the "vice" trade, largely because of the region's proximity and links to the California economy across the border.

Cantú had the support of most of his soldiers and many locals, as well as nearby Americans. As one historian put it, one key to his success "was his ability to project the illusion of order to Americans who had reservations about [their] personal security in Mexico."[33] His government made considerable money by taxing gaming, alcohol, prostitution, and opium, which made up for a lack of financial support from the central government and made possible his success in sustaining troop morale and increasing education, infrastructure, and foreign investment. Cantú also oversaw the highly productive cotton economy in the valleys bordering the United States. Rolland despised Cantú's strongman rule, complex taxes, and use of the U.S. dollar, but it was the lure of the region's financial success under Cantú that helped prompt the Carranza administration to renew attempts to bring the territory under the control of Mexico City.[34] Increasing control over profitable sectors of the economy outside of Mexico City was a key ingredient in the Carranza administration's recipe for defeating its enemies and restructuring the Mexican nation.

Rolland's *Informe* was the most detailed account of local government expenditures and the economy of the Northern District of Baja California ever produced up to that point. He poured his heart and soul into the study, and it remains one of the clearest examples of his passion for the peninsula and his long-standing desire to see it further developed and incorporated into the Mexican nation. Dividing the work into three sections—the economic-administrative situation, problems that demanded the immediate attention of the federal government, and thoughts on the existing local government—Rolland provided a one-hundred-page account of the Northern District of Baja California's taxes, income, expenditures, policies, resources, public works projects, communications, and transportation. He spent another fifty pages providing his personal opinion on the Cantú government and ways to improve the region's growth.

As Rolland admitted, Cantú's administration had made significant strides in education and infrastructure. Cantú had funneled money from taxes into a school construction spree.[35] Rolland was most impressed with a new Mexicali school built with the material dearest to his heart: reinforced concrete. Cantú, Rolland acknowledged, had also made "noble efforts to guard the nationality of the region" through the advancement of transportation and communications networks: roads, especially the Camino Nacional, wire communications from Mexicali to Ensenada, which traversed an extremely inhospitable desert and mountainous terrain, and a number of radio stations.[36] Cantú had developed the area at a more rapid pace than other leaders had in other Mexican territories and states.

Rolland used these brief but possibly genuine compliments to provide a semblance of balance to his overall negative assessment of "Cantú's Kingdom."[37] To make the argument that Cantú was mismanaging the region, Rolland pointed out the wide gap between living standards and development on the U.S. side of the border versus those in Mexico.[38] Development and living standards on the U.S. side were overwhelmingly superior. But as the lower living standards on the Mexican border before and after Cantú's rule show, the roots of the problem went deeper than the issues with Cantú's administration.

Rolland repeatedly contended that Cantú's government was decadent, inefficient, and too favorable to foreigners (somewhat contradicting his statement that Cantú had defended Mexico's sovereignty by building better communications). Most offensive to Rolland's economic sensibility was Cantú's complex system of taxes. To obtain the money for the numerous educational and infrastructure projects, the governor put in place a number of taxes on imports, exports, and the vice trade. Rolland argued that these taxes, despite the improvements in the region, were overly complicated, not well designed, and burdensome. He further contended that the government operated at a "very expensive" level.[39] The taxes and costs went against Rolland's Georgist sensibilities. And of course the single tax was Rolland's prescription for the region.

More than anything else in his report, Rolland's recommendation that the central government increase its presence in northern Baja California had the greatest impact. The area was in many ways still more connected to the U.S. state of California than to the capital of Mexico. Rolland recommended the creation of a permanent federal committee consisting of efficient, intelligent, and loyal administrators to consolidate control in the area. He also pushed the government to increase its military presence by fortifying the region with "two or three thousand men." And despite or perhaps because of the Agua Prieta Revolt, a political rebellion that erupted because of the 1920 presidential election that led to the assassination of Carranza and the rise of Provisional President Adolfo de la Huerta (1920) and President Obregón (1920–24), the central government acted on Rolland's latter recommendation. De la Huerta sent three thousand troops under the command of Gen. Abelardo Rodríguez to face the increasingly noncompliant Cantú, forcing the governor out of office. The dislodging of Cantú surely pleased Rolland, though the subsequent governors, frustratingly for Rolland, did nothing to simplify the tax code or reduce government abuses.[40]

The Agua Prieta Revolt

Rolland had to worry about his own professional and political survival as well. The months leading up to the Agua Prieta Revolt were politically tumultuous. Presidential elections were dangerous. Carranza threw his support behind Ignacio Bonillas, the Massachusetts Institute of Technology–educated engineer, former diplomat, and former minister of communications. Carranza wanted the presidency to go to a civilian and someone who could navigate the rocky relationship with the United States. Bonillas spoke perfect English and had helped mediate the withdrawal of the Punitive Expedition. Carranza may have also considered Bonillas to be someone over whom he could wield significant influence. The problem was that Bonillas was little known among most Mexicans. Few Mexicans knew he even existed until the electoral race. General Obregón, on the other hand, who had lost his arm defeating Villa's forces, had become a war hero and was by far the most prominent military official. He was charismatic and saw himself as the natural candidate to replace Carranza and to unify Mexico. Many Mexicans thought so, too. But these were not the only challengers. Early in the race Generals Pablo González and Alvarado also established themselves as legitimate contenders. Being on the losing side could mean becoming a political outsider or worse.

As his directorship of *El Heraldo de México* shows, Rolland continued to support Alvarado. They had a close relationship dating back to 1914. They were both well-read progressive liberals who shared similar concerns. Rolland had more to gain from an Alvarado presidency. Alvarado prized Rolland's advice and talents. Many people in the know, in both Mexico and the United States, believed that Rolland would become a member of Alvarado's cabinet, most likely the secretary of communications and public works.[41] The relationship between Bonillas and Rolland was less affable. Disagreements between Rolland, Bonillas, and communications official Mario Méndez, another diehard Carrancista, had been what prompted Rolland to resign his position in Carranza's cabinet in early 1915.[42] A González or an Obregón presidency would likely land Rolland somewhere in between the extremes of Bonillas and Alvarado. Obregón and Alvarado had been somewhat bitter allies during their time together in Sonora. Alvarado resented Carranza's promotion of Obregón over him.[43] Rolland was not as close to González or Obregón as he was to Alvarado, but he knew them. Obregón had shown some sympathy toward Rolland's ideas during the Veracruz conference of 1914–15.

The writings of a young telegrapher named Trinidad Flores, who worked as an undercover agent for Obregón during the run-up to the 1920 presidential election and the Agua Prieta Revolt, provide some interesting insights on Rolland. Telegraphers were powerful allies and enemies during the revolution and the rebellions that followed. They exchanged messages and were privy to information. Flores had a favorable opinion of Rolland. According to Flores, Rolland was a practical man who avoided problems. It was Flores who wrote that Rolland had renounced his position as *oficial mayor de comunicaciones* back in 1915 to avoid a prolonged fight with Méndez and Bonillas. Flores also portrayed Rolland as politically astute. Méndez had approached Rolland in 1919 regarding some unnamed articles in *El Heraldo de México*. Rolland, Flores asserted, had convinced Méndez, who was firmly behind Carranza and Bonillas, that he and Rolland were now friends and that Alvarado would not support Obregón over Carranza if Alvarado's quiet efforts to gain the presidency fell apart. Rolland, Flores concluded, was a decent person but was fooling Méndez and playing the different political factions.[44]

On April 23 Obregón, along with Adolfo de la Huerta and Plutarco Elías Calles, his close allies from Sonora, issued the Plan of Agua Prieta, which called for the armed overthrow of Carranza. They argued that Carranza was attempting to impose Bonillas on the country and that the president had used his powers to intimidate Obregón's Sonoran allies, including some of the plan's signatories. Obregón quickly obtained the support of the majority of the military and communications officials, as well as many of the country's governors. He also obtained the backing of peasants, prominent unions, and the remnants of the Zapatistas and Villistas who had never stopped resisting Carranza's rule. Obregón used this coalition to establish himself as the man to unite the opposing factions in Mexico and to bring peace, stability, and prosperity. The insurrectionists took control after only twenty-seven days.⁴⁵ After fleeing Mexico City, Carranza died under mysterious circumstances in the Sierra Norte de Puebla mountains. He was either murdered or committed suicide. Alvarado backed Obregón, siding with his Sonoran colleagues.

Alvarado's support did not go unrewarded. The Plan of Agua Prieta made de la Huerta the provisional president until new official elections would place Obregón in power. De la Huerta made Alvarado the secretary of finance in his provisional government.⁴⁶ One of the first things Alvarado did was to call for the establishment of a free ports system connected by the Tehuantepec Railway. These duty-free trading centers would, Alvarado argued, boost Mexico's economy by increasing trade, industrialization, and the connection of regional markets to the global economy.⁴⁷

Riding the coattails of Alvarado, Rolland would become the head of the free ports project. The general's new position allowed Rolland to bring to the national stage plans they had worked on together in Yucatán. Those plans reflected many of the ideas behind the free ports system. Indeed the Agua Prieta Revolt placed Rolland more at the center of national policy decisions than he had ever been. He never had to fire a shot. Things must have appeared chaotic yet bright to Rolland. Carranza was no longer in the way. The road, or better yet, the ports and railways were wide open.

7

Opportunity, Defeat, and the Death of Virginia Garza de Rolland

ON JULY 29, 1920, as the sun set over the mountains encircling Mexico City, Modesto Rolland sat down for a dinner party at the Restaurant de Chapultepec. The restaurant was elegant. It looked like a gilded gazebo nestled in the park just below Chapultepec Castle. Rolland, along with the directors of the other major newspapers of the capital (*El Demócrata*, *Excélsior*, and *El Universal*), were the honored guests. Representatives of prominent British and French newspapers made the toasts and paid the bill. The guests discussed the new Mexican government and the foreign press in Mexico. The free ports were another hot topic. *El Heraldo de México* promoted the free ports project. After all, Rolland was one of its architects. José Gómez Ugarte, the director of *El Universal* and also attending the dinner, was one of the free ports project's fiercest critics. But such differences did not stop them from sharing an amazing meal: *canapés Muscovite* with *haut sauternes*—a French sweet wine—*consommé madrilène, petites soufflés Walder, foie gras*, mutton, *médaillons de filet aux fonds d'artichaut barigoule* followed by Champagne, *salade de saison* followed by liqueurs, *bombe norvégienne* for dessert, and a round of coffee to top things off.[1] Rolland cared little for alcohol, and even less for Ugarte, but he enjoyed good food and engaging with powerful people.

Rolland had become an important cog in the machine of the revolutionary state. The revolution had been trying but professionally advantageous for him. He had become connected to a broader transnational network of intellectuals, and he was able to implement, even if incompletely, some of his plans for Mexico in Yucatán. He now hoped to seize the opportunity to more fully implement his ideas at the national level.

Despite the decadence of the dinner at the Restaurant de Chapultepec, Rolland was not as exacting in his tastes as some revolutionary leaders, but he obviously wasn't opposed to the finer things in life either. He attended balls that oozed with old Porfirian elitism. He contended that he fought for the common people, but he often dined with the privileged. Rolland had come to the conclusion that to make serious change for all people he had to learn to navigate and participate in the world of wealth and power.

Rolland certainly touted his own level of talent and training, but he often failed to realize his own arrogance. Surely many urban laborers and rural farmers had a hard time trusting revolutionary leaders who spoke for everyday people but dressed like the wealthy. Rolland, like many of his colleagues, argued that they knew what was best, that they were the ones able to bring about real improvement. While eating seasoned goose and carved lamb they designed the lives of peasants. They were top-down planners who rarely listened with any degree of sincerity to the people who, in their view, needed to be raised from ignorance and poverty. Instead they listened to themselves and to their technocratic counterparts in other countries. But despite this arrogance, Rolland would have been right to assume that there were few Mexicans who shared his ability to design massive infrastructure projects and navigate the highest rungs of politics and power. He was one of maybe a few dozen engineers in Mexico with this kind of talent and political savvy. In his view he was doing what had to be done to improve the nation: accomplishing tasks that few people were willing to do or capable of doing.

Between 1920 and 1924, the years in which Adolfo de la Huerta and Álvaro Obregón ruled Mexico, Rolland would have a number of opportunities to put the shovel to dirt, so to speak, on some longtime designs. Rolland acquired new ambitions, too. In addition to becoming the general manager of the Free Ports Commission, he gained a place on the National Agrarian Commission, and he lectured widely and wrote legislation on municipal governance. He also became a leading figure in the promotion of a new communications medium—radio broadcasting. Through radio Rolland imagined the greatest minds in Mexico, perhaps in the world, beaming into the homes of impoverished, illiterate laborers and isolated Indians in remote villages. Specialists would talk. The ignorant would listen. Rural *campesinos* would become more civilized. The nation would be brought together. For Rolland the potential of radio broadcasting as a force of education and unification could not be overstated.

But opportunity does not guarantee success. Rolland found that despite his talents it was the support or lack of support from the president that all too often determined the outcome of his endeavors. His long-held visions could be cut down instantly by political whim. Throughout his life Rolland argued he was apolitical. He stated the point over and over again, and perhaps he believed it. He absolutely believed there was a difference in being a politician versus being an engineer who had to navigate politics (even though some of his engineering colleagues became career politicians). By the 1920s, though, Rolland was an experienced propagandist. He possessed strong nationalist sentiments, clear progressive tendencies, and an obsession with the ideas of Henry George, ideas he had to promote in a political field full of other intellectual influences, ranging from Marx to more traditional liberal thought. It is hard to see how Rolland could not have realized just how political he had become. Whether he believed his own rhetoric or not, politics continued to play a crucial role in the outcomes of the projects he directed and the course of development in Mexico more generally. As soon as he had reached high-level positions and had begun to carry out projects of national and even international significance, backroom political decisions often threw his endeavors into disarray.

Rising to National Prominence

The Congress elected de la Huerta provisional president on May 24, 1920. The following day Rolland's *El Heraldo de México* praised the new head of state. On the same page of the newspaper, in a smaller article off to the side, Rolland and his staff eulogized Carranza. The newspaper had been critical of Carranza in the months leading up to his death. Nonetheless, the staff argued, Carranza needed to be recognized for his contributions to the revolution, not just his faults.[2] Rolland acknowledged Carranza's importance to the establishment of the new revolutionary state, but he did not linger long in his mourning of the man he condemned as too conservative, even though he had once vociferously defended and praised him. There was progress yet to be made.

Rolland obtained two important positions during de la Huerta's brief tenure: he became general manager of the Free Ports Commission and a member of the National Agrarian Commission. The posts reflected a natural progression from his roles in Yucatán, and he obtained both positions through Alvarado's own ascendancy. Of course it was unclear whether Obregón, once elected, would keep in place the people put into top government positions under de la Huerta. Alvarado and Obregón had a long, competitive, and not always friendly relationship.[3] With the election looming—de la Huerta's term would not even last all of 1920—there was a dash to accomplish meaningful policies, to rush them through. The first thing Alvarado pushed for as secretary of finance was the creation of the free ports.

The idea to build a canal or railway across the Tehuantepec isthmus to facilitate Mexican and global trade was an old one. In the sixteenth century the conquistador Hernán Cortés had explored the region and believed it could provide a path to connect Atlantic and Pacific trade. Nineteenth-century French and American politicians and entrepreneurs had considered Tehuantepec one of a few places where a canal linking the two oceans was feasible. Panama, however, won out, becoming the site of the only transoceanic canal in the Western Hemisphere. It was completed by American, Panamanian, and Caribbean laborers in 1914. Desiring to improve Mexico's economy while lessening the influence of the United States, Porfirio Díaz had partnered with the British engineer and business magnate Sir Weetman Pearson to complete a functional railway across Tehuantepec and to construct efficient ports at each terminus. The most expensive project ever undertaken by the Díaz administration, the project was completed by the Pearson and Son firm in 1907. It was amazingly successful. By 1913 trains were carrying more than 850,000 tons of goods annually from one port to the other. Much of this business came from sugar producers in Hawaii. The following year, however, traffic went into steep decline. The reason was a threefold storm of major events: the completion of the Panama Canal, World War I, and the escalation of the Mexican Revolution.[4]

The constitutional congress dissolved the contract between the government and Pearson in October 1918. Pearson, now Lord Cowdray, was happy to get out of it. He had become more dedicated to his oil concerns, and the government paid him full indemnity.[5] The debate after 1918, of which Cabrera, Rolland, and Alvarado were very much a part, was how best to bring the railway and ports back up to their 1913 level of success.

Turning Puerto México and Salina Cruz into free ports was one of the more popular ideas. Cabrera reached out to officials in Copenhagen and to Mexican and U.S. technocrats, politicians, and business operators, including Fernando González Roa, Mariano Cabrera, Lorenzo Pérez Castro, Carlos M. Hammeken, and William H. Ellis, to explore how to move forward on the idea.[6] Rolland also began a more in-depth study of free ports. Alvarado and de la Huerta decreed the free ports into existence, which the Congress, after a brief but serious public debate, signed off on.

The free ports themselves were not a new concept. Arguably their roots go back as far as ancient Tyre and Carthage in the Mediterranean and early modern Genoa. Their modern successors, those that directly influenced the free ports plans in Mexico, had been established in the late 1800s in northwestern Europe, in Hamburg, Bremen, and Copenhagen, among other cities. The recently unified German state was their most ardent promoter, and Hamburg was far and away the most successful of them.[7]

The definition of what constitutes a free port has varied over time and according to the legislators of their host countries. In general, though, a free port was (and is) a policed area separated from the rest of a port town and customs oversight.[8] In the free port all ships could unload and load goods, and entrepreneurs could store materials, manufacture products, and reexport goods without regular customs costs and formalities. If traders carried their goods out of the free ports and into the host countries, then they would have to undergo customs checks and pay duties. Free ports were thus partially denationalized tax-free or tax-reduced trade zones used to entice international business.

Alvarado's free ports declaration sparked immediate debate. Supporters argued that the project would benefit Tehuantepec, improve ports, and improve the Mexican economy by drawing in entrepreneurs and creating new places of industrialization. Proponents also stressed Mexico's strategic location; it neighbored the United States and was centered between the markets of Europe and Asia.[9] Rolland further contended that the free ports would help defend Mexico against its increasingly powerful northern neighbor. They would provide foreigners, mostly Americans, with access to Tehuantepec, cheap labor, and tropical resources but at specific, supervised locations policed by Mexicans. Rolland feared that Tehuantepec would become too strategically important as a trade route for powerful international capitalists to ignore. If Mexico did not develop the region for itself and for those foreign capitalists, under Mexican terms, powerful foreigners might attempt to take control of the isthmus.[10]

Critics levied a number of concerns. Some industrialists feared they would suffer because traders in the free ports did not face the same taxes. Smugglers had taken advantage of free-trade zones along the U.S. border to smuggle goods, undermining legal businesses. Opponents of the free ports wondered if Mexico's limited consumption of Asian goods was sufficient to merit establishing a free port in Salina Cruz. They argued further that European mercantile interests would continue to prefer the Panama Canal over Tehuantepec. The project would, they contended, lose money, waste time, and possibly hurt Mexican businesses.[11] And what about the people who lived in Salina Cruz and Puerto México? Although proponents promised new wealth and opportunity, people with homes and businesses in the area zoned for the free ports would have to move. Vendors in the towns were for the most part banned from selling goods to people in the free ports. Surely they questioned the value of the promised progress, if they were informed at all.[12]

Rolland became the main spokesperson for the project. He sent articles to newspapers and spoke at forums. He felt obligated to "clarify unclear points" and "obligated yet again to fight" for Mexicans, to convince them of the value of this endeavor so essential to reconstruction. He emphasized that the ports would bring prosperity to all people, that they were for the "children, for those who struggle . . . [as well as] commercial captains, bankers, and intellectuals."[13] In a bid to gain greater backing from industrialists and laborers alike, Rolland strove to clarify that he was not inherently opposed to capitalism; he instead supported a nationalist, socially conscious form of capitalism.

Despite merciless criticism from *El Universal*, the project continued. The *Diario Oficial* published de la Huerta's decree establishing the ports on October 11, 1920, and the Congress gave its stamp of approval on November 19.[14] The Ministry of Finance would "organize and direct the free ports," but the "organization, administration, and management" would be "directly under the control of a council composed of five members," one of whom would be the chair. De la Huerta and Alvarado tasked Rolland with heading the project. The Ministry of Finance received $500,000 to start the required expropriations, construction, maintenance, and hiring of personnel.[15]

Rolland also became a member of the National Agrarian Commission in the summer of 1920. The nine-member commission, five of whom, according to law, had to be engineers, represented the various ministries and different political interests that had supported the Agua Prieta Revolt. Rolland represented Alvarado and the Ministry of Finance.[16] Rolland was put in charge of "conflicts created in Yucatán, Baja California, Campeche, and

Veracruz."[17] The new political victors counted among their supporters former followers of Emiliano Zapata and former Carrancistas and Villistas, all of whom called for more radical agrarian reform policies. The de la Huerta administration pushed through a law that allowed for the creation of *ejidos*—the Ley de Tierras Ociosas, or Law of Unused Lands (June 23, 1920), which allowed municipal authorities to temporarily give permission to people in need to rent and to farm unused lands held by large private interests. De la Huerta made Antonio I. Villarreal, who had been a progressive governor of Conventionalist-controlled Nuevo León in the last months of 1914, the new secretary of agriculture and chair of the National Agrarian Commission. Rolland represented the Carrancista engineers who promoted greater land redistribution in a vein similar to what had been attempted in Yucatán. Lastly, there were members, including Andrés Molina Enríquez, who had been aligned with Huerta briefly but who had been brought back into the fold because of their friendships with top leaders and their importance in shaping Mexico's revolutionary agrarian policies.[18]

Rolland remained active in other prominent organizations as well, most of which were connected to the Engineers' Center. He helped found the Club de Estudios Económicos-Sociales, or Social-Economic Studies Club, which hosted roundtables and gave professional presentations on myriad topics, many of them focusing heavily on the protection and exploitation of natural resources. Carrying on their tradition from the last days of the Díaz administration, Rolland and the other members claimed to be pro-Mexico but nonpolitical. They wanted only to "study all problems of a national character ... the petroleum question, municipal autonomy, the agrarian problem, cadastre, taxes, etc."[19]

The group consisted of approximately seventy-five members, mostly influential engineers but also lawyers, labor leaders, business operators, politicians, and generals. Some of the prominent members included Manuel Gómez Morín, who was a Ministry of Finance official (and future founder of the Partido Acción Nacional, or National Action Party); Roque Estrada Reynoso, a member of the Congress; and engineers Ignacio Díaz Soto y Gama, Mariano Cabrera, and Edmundo Cardineault. Gen. Eduardo Hay and Gen. Ramón F. Iturbide were also members, as were the archaeologist Alfonso Caso and labor leader Vicente Lombardo Toledano. The group elected Rolland as its president, showing the widespread regard for his work ethic, abilities, and influence.[20] They shared similar visions of a more progressive Mexico, though they differed in how to address important issues such as taxation, agrarian reform, and labor.

Rolland continued to publish prolifically. In addition to founding the journal *El Hombre*, which was another Georgist, progressive-style periodical, he wrote *El desastre municipal* (The municipal disaster), a two-hundred-page book that received significant attention among revolutionary leaders interested in city planning. The book focused on the need for Mexico to develop better-governed communities. He was especially concerned with elections, the division of labor, and taxes. For all his work on land reform, Rolland was a fan of urbanity. Cities, to Rolland, represented the future and the key to modernity. He believed they were good for democracy, collaboration, and the improvement of humanity. The city was the basis of civilization and "all its manifestations: education, arts, culture, industry . . . the brain and the machine of excellence." Cities were, Rolland continued, "the heart and the central nervous system of the actual world."[21] He called for collaboration between classes in order to build modern, freer cities. Echoing progressives of earlier decades and modernist city planners in Europe and the United States, Rolland concluded, "The city is rapidly transforming and no longer the dreaded object of the patriarchal life. For the first time in the history of humanity, all acquisitions of science, industry, etc., lend their support to the social direction and the rule of the people over their own lives."[22]

To continue the improvement of urban spaces, Rolland argued for the creation of serious civil service laws, municipalities governed by commission, and the incorporation of referendums or recall elections. Yet again he called for the simplification of taxes. He condemned the Mexican revolutionary leadership for doing too little to change exploitative and tributary processes that concentrated political power in the hands of *caudillos* and the Mexican president, who doled out municipal posts as political favors.[23] For positive examples of town governance and planning he looked to the leaders of Western industrial societies, particularly Germany, the United States, Britain, and France. He praised libraries, university-trained city planners, civic- and city-minded elected officials, and well-organized communications networks. Rolland's dream city, as depicted in *El desastre municipal*, was essentially of German design, with French and English gardens and U.S. progressive-style governance.[24] Rolland later told a Georgist convention in 1930 that in order to write the book he had "spent a part of the patrimony of my children; who will forgive me when they shall understand how ardently their father has worked to create a better country for their future use."[25] Rolland undoubtedly cared for his country, and there is also no doubt that Rolland's family suffered because of his obsessive need to improve Mexico.

Dreams, Difficulties, and Death

The transition to the Obregón government in December initially brought little change to Rolland's projects. Along with Alvarado, the engineer Norberto Domínguez, and the lawyer Ezequiel A. Chávez, Rolland drew up new legislation to establish a civil service law, which he had already laid out in *El desastre municipal*.[26] He continued as head of the free ports, as a member of the National Agrarian Commission, and as a prominent public figure in general. Obregón signed off on the ejidos law (the Ley de Tierras Ociosas) only eight days after his inauguration, renewing the granting of communal lands to wronged villages, while allowing more autonomy at the state and municipal levels. But there were serious problems with establishing who was eligible for land restitution and getting the expanding bureaucracy to operate at a functional level, a problem that had plagued the commission since its establishment in 1916. Rolland supported the law, especially its federalist-style approach, but it was ultimately rejected by Obregón, who wanted more control over land distribution. Instead the president signed off on another bill that provided more central control the following year.[27]

Rolland appears to have been involved in almost every conference on development held in Mexico City in 1921 and 1922. He helped organize the Primer Congreso Nacional de Caminos, or First National Congress on Roads, which was held at the National School of Engineering in September 1921. He was also a vocal participant in the highly contentious inaugural Congreso Nacional Agronómico, or National Agronomic Congress, where he called once again for the nationalization of land and pushed for the scientific and sustainable exploitation of forests and other resources.[28] Both congresses were part of the September 1921 events commemorating the centennial of Mexico's formal independence from Spain. Obregón painted his new administration and the centennial events as exhibiting a new, cooperative era of Mexican independence, equity, and modernity. There may have been more democratic debate, but division often trumped cooperation. The following April saw Rolland participating in the Segundo Congreso Nacional de Ayuntamientos, or Second National Congress on Town Councils, where he continued to push for Wisconsin-style townships.[29] In November 1922 he organized the first Convención Nacional de Ingenieros, or National Convention of Engineers. Meeting in the Engineers' Center, its members focused on city planning, municipal governance, and unionization. They set up a space in which to share information through publications, public talks, and films. They also established plans to do outreach work in small towns and to train individuals from those towns in methods of modern municipal democracy and design.[30]

The differences displayed during the centennial conferences exemplified the difficulties in bringing the various factions of the revolution to terms over land policies and other issues of reconstruction and development. Agrarian reform had become a hugely political affair. Many of Obregón's strongest supporters—members of the Congress and his new cabinet—wanted profound changes. Former followers of Zapata who were prominent in the influential Partido Nacional Agrarista (PNA), or National Agrarianist Party, including Antonio Díaz Soto y Gama, demanded more restitution for village ejido lands. Many members of the Partido Liberal Constitucional (PLC), or Liberal Constitutional Party, including former Constitutionalists such as Gen. Enrique Estrada and Rafael Zubarán Capmany, called for more land redistribution but supported various small private-property schemes. Molina Enríquez, a member of the PNA until 1923, championed medium-sized family farms, as he had in *Los grandes problemas nacionales* in 1909. He also promoted the federal government's right to own the land and resources, justifying his stance not with socialist or Georgist doctrines but laws from the colonial era.[31] Rolland represented a small minority of former Carranza-linked officials who promoted Georgist policies. Having been handed a near impossible job, in disagreement, and facing constant attacks from the Congress, the press, and eventually Obregón, Villarreal and the rest of the National Agrarian Commission moved forward.[32]

Rolland became quickly frustrated with the National Agrarian Commission. The commission made him head of the Executive Committee of Ejido Administration in late September 1921, but he complained that Obregón and Villarreal, though increasingly at odds with each other, both used ejidos as a political tool.[33] Villarreal resigned amid criticism from the president and corruption charges. Rolland meanwhile attempted to move the commission in a more Georgist direction, shifting away from ejido distribution and toward individual plots and ultimate state ownership of the land, something that Molina Enríquez backed as well. But first Rolland called for a slowdown in the granting of ejidos in order to better organize those that had been distributed and to better comprehend changes in production.[34] Formerly criticized by Carranza for too hastily distributing lands, Rolland now found himself facing charges of conservatism and slowing redistribution.

Obregón too was perturbed with the National Agrarian Commission and other members of his new coalition who refused to follow his agenda. After forcing Villarreal out, Obregón "accepted the resignation" of Rolland and other members of the National Agrarian Commission in February 1922. Rolland was genuinely surprised by his ouster. He wrote personally to the president, pleading for an explanation.[35] Obregón justified his decision by saying that Rolland needed to focus on his duties as manager of the free ports.[36] The ousting of Rolland from the National Agrarian Commission was really part of a larger political shift. Zubarán, then serving as secretary of industry and commerce, resigned along with two other PLC cabinet members over Obregón's rejection of their party's push to strengthen the Congress and weaken the executive. Rolland was not a member of the PLC, but he was closely associated with many of its members.[37]

Obregón had no intention of letting anyone weaken his authority. The state was fragile. He was not going to leave the fate of the nation to the constantly bickering Congress. He would use land donations as a way of strengthening the presidency, stabilizing the state apparatus, pacifying the country, and securing his own power. The president showed little patience for drawn-out debates about agrarian policies, economics, or legislative power.

Rolland attempted to direct his attention to less politically contentious issues in hopes of continuing some form of progress. He became enmeshed in the development of a new means of communication—radio broadcasting.[38] The newest form of electronic communication, broadcasting could reach thousands, it not millions, of people. Rolland, as well as others who believed in the medium, saw the technology as having immense potential to unify and "civilize" the nation. It would bring intellectuals, teachers, history, and the best music to the most distant villages.[39]

There was a small but growing population of radio enthusiasts throughout the country. Some of them had been experimenting with radio since before the revolution. The war had expanded the use of radiotelegraphy among the military factions but had dampened the use of the medium by aficionados. Possessing radio equipment had become dangerous because militants often saw owners as potential spies. But Obregón's presidency brought a brief renaissance in radio use among hobbyists. Enthusiasts, often from middle- and upper-class families, began experimenting with some regularity, talking with each other and reaching out to aficionados in other parts of the world, especially the United States.[40]

Rolland helped form the most powerful radio lobby in the country, the Liga Central Mexicana de Radio (LCMR), or Central Mexican Radio League. Founded in early 1923, it was the result of the merger of the first Mexican radio society, the National Radio League, established on July 6, 1922, the Central Radio Club, and the influential Engineers' Center.[41] The new group elected Rolland as its president, and it quickly became crucial in promoting broadcasting in Mexico. The LCMR drew up the first regulations for the medium, which Obregón signed off on with some caveats.

The LCMR-written regulations drafted between March and September 1923 provided a compromise between the state and businesses, and the hobbyists. The guidelines established rules for licensing and taxation (the latter included by Obregón, much to Rolland's displeasure). The rules allowed for private profit, but the Ministry of Communications and Public Works had to approve all permits, for which businesses had to pay an annual fee.[42]

Most LCMR members stressed elitist visions. The 1923 regulations favored commercial radio operators, but a number of LCMR members lobbied the government to take a greater role in using the medium as an educational tool. Rolland and Manuel M. Stampa, a prominent experimenter and officer of the league, declared that the LCMR's purpose was to coordinate the propagation of radio, which "suddenly puts the men of remote villages in contact with the civilization of the most advanced centers of culture."[43] When Rolland protested the tax put on radios, he argued that the fee impeded the spread of receivers, which college classrooms and households could use to access transmissions of important conferences and lectures, allowing Mexicans to build a shared sense of nationality and a respect for science and the creations of high culture.[44]

Shortly after the opening of the first commercial broadcasting station, CYL, which was owned by *El Universal* and Raúl and Luis Azcárraga—proprietors of radio store La Casa del Radio and brothers to Radio Corporation of America (RCA) sales agent and future media magnate Emilio Azcárraga—Rolland helped organize Mexico City's Grand Radio Fair. It was an impressive exhibit of broadcasting and radio products. Lasting from June 16 to 25, booths displayed locally made radios and the latest U.S. receivers. Participants visited elaborate displays. In one of the wilder ones women gave away tobacco company El Buen Tono's "Radio" cigarettes while wearing mock antenna hats. The company was in the process of constructing the country's second commercial station, CYB, on which it would continue to promote its cigarettes.[45] Rolland continued to campaign for the medium during the remainder of Obregón's presidency, especially pushing its potential to educate and unify the nation.

In addition to developing radio, Rolland poured attention into the free ports. He made significant progress in building and improving the facilities. In July 1923 Rolland once again returned to New York, this time to promote the free ports project, which was now well under way. He coordinated closely with R. A. C. Smith, a former New York City dock commissioner, and Carlos A. Félix, Obregón's "financial agent" in New York. At a luncheon with members of the U.S. Chamber of Commerce, the Merchants' Association, and the New York Board of Trade and Transportation, Rolland put aside his criticism of foreign capitalists. Instead he called for closer economic relations.[46] He promoted Mexico's "wealth of labor[,] from 85 cents to a dollar a day," and access to Mexican raw materials. He called on Americans to use the ports to make manufactured goods out of imported and local raw materials. The ports, Rolland declared, provided the most direct route to Oceania, Asia, and the west coast of South America. Other members of the Mexican delegation, however, assured U.S. business operators that Tehuantepec would not compete with the Panama Canal; the operations would work in harmony, with Tehuantepec helping to offset the congestion already developing in the waterway in Panama. Rolland told the crowd that the Mexican government would make some revenue from renting the warehouses—there would also be a docking fee—but that the goal of the ports was not to produce revenue but to encourage more commercial relations.[47]

Rolland and the rest of the executive board had worked with private steamship companies, including the Naviera Mexicana del Golfo (Mexican Shipping of the Gulf) and the Naviera Mexican del Pacifico (Mexican Shipping of the Pacific) to build a sort of merchant marine that helped transport goods between the free ports, other Mexican ports, and San Pedro (close to Los Angeles), San Francisco, and New Orleans. There were seven ships for the Pacific and five for the Gulf of Mexico. The services, as initially set up, were heavily tilted toward wooing U.S. participation.[48]

By the summer of 1923 the Obregón administration had already sunk hundreds of thousands of pesos into the project. It had created an inland "free port" in the village of Rincón Antonio, approximately halfway between Puerto México and Salina Cruz. It was to serve as a midway storage facility and duty-free industrial center.[49] The Sonoran city of Guaymas was also to be included, likely as an incentive to President Obregón, who hailed from the area. Laborers had started fencing off the free ports, repairing dilapidated warehouses, modernizing the wharves, and making the port at Salina Cruz, which was an artificially constructed port, navigable. Salina Cruz took near-constant dredging and was, in many ways, less than an ideal location for a port, something the Pearson firm had realized after creating it. The ocean constantly swept in sand, filling in dredged channels.[50]

While in New York, Rolland also attempted to salvage his marriage. The marriage had never been an easy one. Strained by the revolution, moving, and multiple abandonments, Virginia Garza de Rolland and Modesto Rolland's marriage completely fell apart after the death of their youngest daughter, Carmelita, who had died after falling into an open well. One factor pressuring Rolland to rebuild his marriage was the disfavor of political colleagues and superiors. De la Huerta, now secretary of finance, supposedly told Rolland to get his family life in order.[51] Promised a nice home, private schools for the kids, and financial security, Garza de Rolland and the children returned to Mexico City.[52]

Only months later, on the afternoon of December 6, the Rolland family's chauffeur drove Garza de Rolland and Enriqueta, Virginia and Modesto's first-born child, from their house to go shopping in the city, possibly for Enriqueta's upcoming *fiesta de quince años* or *quinceañera*. She was only days away from that celebration of becoming a woman, a rite of passage marked by her fifteenth birthday. A train approached on an upcoming railway crossing. The chauffeur, possibly unaware, continued and the train plowed into the car. The intense impact ejected Enriqueta, but amazingly she lived. Garza de Rolland was less fortunate. The train's force threw her out of the car and into a tree, killing her instantly. Her death sent shockwaves through Mexico City's social circles. Newspaper editors immediately took to their typewriters to spell out their condolences to Rolland, their esteemed colleague. At that point no one was sure what exactly had happened. The newspaper *El Mundo* reported that the chauffeur's carelessness had caused the accident and that the unobservant chauffeur, who survived the wreck, had fled in fear of the police, who were searching for him.[53]

The Adolfo de la Huerta Rebellion

Rarely does death come at a good time for families, but Garza de Rolland's death may have saved Rolland from joining the wrong side of a new rebellion. Campaigns for the next presidential election were well under way by December. The top two contenders were Adolfo de la Huerta and Plutarco Elías Calles.

Obregón favored Calles. Although all three of these men had been close at times—they all hailed from Sonora and had together planned the Agua Prieta Revolt—de la Huerta was outraged that Obregón chose Calles over him. Most members of the PLC and a number of Rolland's closest colleagues, including Alvarado, Zubarán, Martín Luis Gúzman, and Villarreal, sided with de la Huerta.⁵⁴ Asserting that Obregón was imposing Calles on the nation and that the election would not be honest, de la Huerta and his supporters proclaimed rebellion from the port of Veracruz on December 7, the day after Garza de Rolland's death.⁵⁵ Rolland stayed home. Despite his rocky relationship with Garza de Rolland, her death was difficult for Rolland. His children were now under his direct care. It was also around this time that Rolland started caring for his sister Victoria, who suffered from a disabling mental illness. He would continue to take care of her at his Mexico City house until he sold it in the late 1950s.⁵⁶

The De la Huerta Rebellion posed a serious threat to the Obregón government. It exploded into a full-scale civil war, with nearly half the military rising against the president. Rebels controlled many of the important ports and radiotelegraph stations, and the Delahuertistas' initial drive on the capital caught the government off guard. Many residents believed Obregón would abandon the capital in order to fight from another location.⁵⁷ The revolt, however, stalled after distrust and disagreements among the leadership halted the advance on Mexico City, which allowed Obregón to launch a successful counterattack.⁵⁸

Rolland never stated why he did not support the rebellion. He was upset with Obregón's heavy-handedness, his handling of ejidos, and his taxation of radio technology. Many if not most of the powerful people Rolland called friends had joined the insurrection. Not having Garza de Rolland around to be the main caretaker likely left him somewhat confounded and uncomfortable. But to venture a little further into the world of speculation, there are other likely reasons for Rolland's silence and then ultimate rejection of the insurrection. For one, we know that Rolland was not fond of revolution. A man of letters and blueprints, he was careful to distance himself from violence during political upheavals and had worked carefully to stress his nonpolitical nature and to balance his praise and criticism of Carranza and Obregón. He had also invested serious time into the free ports, municipal governance, and radio development. He had been working tirelessly on sharing ideas about reconstruction at the Engineers' Center. Rolland wanted to build Mexico. Further rebellions would do little to aid that cause. His allegiance tended to be influenced most by whomever and whatever could bring about stability and help him achieve his own goals to improve the country. Perhaps he thought the De la Huerta Rebellion was too risky and too destructive.

Despite the death of his wife and the insurrection of colleagues, Rolland did not sit out the rebellion completely. Remaining loyal to the Obregón administration, and more so to his work, Rolland became a middle man, helping radio consumers, aficionados, and commercial stations navigate wartime restrictions on the medium. The brief democratic opening in radio that had occurred in 1923 snapped shut. The military closed down one station because of its support of de la Huerta. The Obregón government stopped all sales of transmitters and required people to register their equipment. Rolland helped the Obregón administration distribute new emergency guidelines by speaking with the press and giving talks through the LCMR and other public forums. He also worked to keep the nascent radio industry as open as possible in the face of these authoritarian measures.[59]

As the De la Huerta Rebellion petered out, Rolland declared the free ports open for business on July 21, 1924. He had already received requests from American companies for warehouse space. There were also companies interested in renting room for manufacturing.[60] In the days leading up to the free ports' formal inauguration, which occurred in Salina Cruz, he led a delegation of American, French, and German representatives and press agents on a tour of the railway and ports. According to one U.S. journalist, the port at Puerto México was in good shape, and though the warehouses were in poor condition, they were in the process of being repaired. He called Salina Cruz "deserted," stating that the sand-choked channels were still an issue, but he relayed, with a sense of astonishment, that the warehouses, cranes, and dry dock were in good shape. They, along with free port at Rincón Antonio, were "carefully screened with barbed wire from the towns," and "a special police force" had been hired for security.[61] Yet, even with the generally positive assessment by reporters, the embers of the Mexican Revolution and the De la Huerta Rebellion still smoldered somewhere in the minds of foreign business operators, causing hesitancy about investing in Mexico. However, for the moment, there was relative peace and things looked to be back on track for the Mexican state and Rolland.

Then, without justification, Obregón shut down the free ports. It devastated Rolland. The president had again crushed his aspirations. Rolland believed the decision to shut down the ports was the result of pressure from powerful American interests influencing the Bucareli Accords of 1923, which resulted in U.S. official recognition of Obregón's government. Rolland specifically blamed U.S. interests involved with the Panama Canal.[62] Sixteen years later, after a new government had placed him in charge of bringing the free ports back from the dead, Rolland wrote that the sudden cancellation in 1924 "had been a strong blow to the Isthmus of Tehuantepec. The operation never had a chance to prove itself." The government had "provoked the natural distrust of the international business that was to use the free ports of Salina Cruz and Puerto México."[63] To Rolland, the action was an inexcusable "wrong against the country."[64] He became immensely bitter and less certain about the revolutionary leadership. Coming only months after Obregón had ousted Rolland from the National Agrarian Commission—a move Obregón had justified by telling Rolland that his attention needed to focus more on the free ports—the president's backtracking crushed Rolland. He had seen his greatest ambitions dashed with a stroke of the presidential pen. The early 1920s had been an immense roller coaster for Rolland. Indignant but determined, he pressed on.

8

A Stadium for Stridentopolis

IN APRIL 1923 MANUEL Maples Arce spoke the first words broadcast over a Mexican commercial radio station. He wrote a poem just for the occasion: "TSH (El poema de la radiofonía)" (Wireless [a poem about radio broadcasting]).

> Stars launch their programs at nighttime over silent cliffs. Words, forgotten, are now lost in the reverie of an inverse audion. Wireless telephony like footsteps imprinted in the empty shade of gardens. The clock of the mercury moon has barked the time to the four horizons. Solitude is a balcony open onto night. Where is the nest of the mechanical song? The insomniac antennas of memory receive wireless messages of some disheveled farewells through sleepless antennas. Shipwrecked women lost in transatlantic ambiguity; their cries for help burst like flowers on the wires of international pentagrams. My heart drowns in the distance. And now a "Jazz-Band" from New York; vice blossoms and engines thrust in synchronized seaports. Nuthouse of Hertz, Marconi, and Edison! A phonetic brain shuffles the accidental perspective of language. Hallo! A golden star has fallen into the sea.[1]

Maples Arce was a young lawyer-in-training who dressed in dapper suits and wrote radical iconoclastic poetry celebrating all things modern.[2] He was an *estridentista*, or Stridentist, which is what he termed the artistic movement he founded. The Stridentists were a small but influential group of avant-garde writers and artists. Maples Arce delivered "TSH (El poema de la radiofonía)" in the days leading up to Mexico City's Grand Radio Fair of June 1923. Modesto Rolland, who oversaw much of the planning for the fair and was a well-known figure in Mexico City's radio and social scenes, surely listened to this inaugural broadcast. He was one of the first people to own a good radio receiver in the capital. In fact Rolland had a room in his Mexico City residence specifically designated for radio equipment. He, like Maples Arce, loved radio—its modernity, its potential, its power.[3]

It would not be until two years later that the worlds of Rolland and Maples Arce would more directly meet. It was not their shared passion for radio that drew them into closer circles; it was the administration of Gen. Heriberto Jara, who was serving as governor of Veracruz. The preceding governor, Adalberto Tejada Olivares, had taken a post in the cabinet of Pres. Plutarco Elías Calles. Tejada had briefly attended the National School of Engineering when Rolland was a student but dropped out due to financial hardship.[4] After obtaining his law degree, Maples Arce moved in the other direction, leaving Mexico City for Veracruz to become a judge in Xalapa, the state's picturesque capital. Jara, impressed with Maples Arce's integrity, subsequently made him his secretary of internal affairs.[5] While Maples Arce was a judge, Jara hired Rolland to design and build a stadium, a showcase of the modernizing drive of the Jara administration and something that could itself be used for other celebratory displays of modernity and progress.

Rolland never considered himself a Stridentist; he came from an older generation than most of these iconoclastic poets and painters. But his stadium became a pillar of the Stridentist aesthetic in Xalapa, where a number of the most prominent writers and artists of the movement had relocated. These futurism-inspired artists envisioned Xalapa as Estridentópolis, or Stridentopolis, a metropolitan, modern, industrial center built of concrete and radio stations. Their projection of Stridentopolis was far more technologically advanced and industrial than Xalapa actually was. Aureliano Hernández Palacios, a Xalapa artist who was a student in his teens during the 1920s, later reminisced that in that decade "Xalapa was quieter, with a provincial spirit, more tranquil and more frequently and demurely huddled under a shawl of mist."[6] But perhaps the Stridentist vision was not completely egregious. Rolland would build a new stadium, one that unabashedly celebrated its reinforced concrete architecture, and Jara had ordered the construction of a radio station, a dam, and a number of buildings and roads. Pavement and cars had made their way to the provincial town, and gas stations had begun to spring up. There was a punctual rail service that connected Xalapa with surrounding villages and the port of Veracruz.[7]

Rolland and the Stridentists worked in Xalapa at the same time—during the presidency of Calles. Rolland and Calles did not have a close relationship. Álvaro Obregón had put in the initial order for the closure of the free ports, but it was Calles who nailed the coffin shut. Calles appears to have had no interest in working with Rolland.[8] In addition, the fact that many of Rolland's closest colleagues had been members of the De la Huerta Rebellion surely did little to instill a sense of trust between Rolland and the new president. Jara and Calles had a functional relationship, but they too were mistrustful of each other despite the fact that Calles backed Jara's gubernatorial run. Jara, alongside Tejada, had provided military leadership in Veracruz for Calles and Obregón during the De la Huerta Rebellion, but Jara had formerly been friends with some of the rebellion's leaders. Calles and Jara's relationship grew even tenser during Jara's years as governor due to disagreements over state autonomy, labor issues, and Jara's demands on oil companies.[9]

Jara and Rolland, on the other hand, shared many of the same political acquaintances and had established a close relationship since Jara's collaboration with Alvarado during their military invasion of Yucatán in early 1915.[10] This relationship continued during Rolland's time working with Alvarado. Jara and Alvarado had differed on military strategy, but Jara was an admirer of Alvarado as a person, calling him a man of "high ideals" who fought for social justice.[11] Jara was aware of the ideas and projects that Rolland had brought to Alvarado's Yucatán.

Hoping to build on the initiatives started in Yucatán a decade earlier, Jara invited Rolland to Xalapa, providing him political refuge and new opportunities.

Their relationship exhibits how Rolland's ability to adapt and thrive during different presidential administrations was tied to personal connections in one of the multiple cliques existing in the national bureaucracy and in different state governments. Rolland again found refuge among alliances made at the Veracruz conference of 1914–15. Many of these allies had become powerful leaders; in the case of Alvarado and Jara they had become governors of states. Rolland used these connections to find safe havens during times of political opposition in the capital. There were pull factors involved, too. Jara, much like Alvarado, offered Rolland new opportunities to make his visions come to life. There would be fewer bureaucratic and political difficulties at the state level. Jara granted Rolland significant freedom to design a major public works project and to reenvision a large section of Xalapa.

Rolland's professional wanderings further exhibit just how difficult it was for him to bring large projects to completion without politicians undermining them. By the time Rolland found himself building a stadium in Xalapa in 1925, he had seen his projects for land reform at the regional and national levels wax and wane as political leaders came and went. He found strong government support for his free port designs only to have them crushed as soon as they got off the drawing board. President Calles further quashed Rolland's plans for the ports and connecting infrastructure in Tehuantepec, keeping Rolland out of the federal government altogether. There was a lack of consistency in planning at the top levels. During the tumultuous years of the late 1910s and the 1920s, personal and situational politics trumped consistent support for long-term development projects, which suffered from major shifts in policies and a high turnover rate among government planners and engineers, who struggled to survive by finding new positions in the states, the private sector, or within different federal agencies.

Jara, for his part, was bent on making a lasting legacy in Veracruz. He spent lavishly on public works projects and incorporated a younger generation of thinkers, including the Stridentists, into his government. The spending spree caused severe deficits for the government of Veracruz but allowed Rolland and the Stridentists to work under a benefactor who allowed them to complete significant projects and advance their notions of progress. The stadium Rolland built, the Estadio Heriberto Jara Corona—better known as the Estadio Xalapeño or Xalapa Stadium—stands to this day, a clear testament to its design and construction.[12] The debts, on the other hand, would undermine Jara's rule and darken any visions of more comprehensive, progressive city development. Big plans have big costs, a prohibitive reality that would continue to slam into Rolland's dreams.

Xalapa

The state of Veracruz had grown on Rolland. He had crisscrossed its territory during a number of trips during the prior decade. He particularly liked the Sierra Madre Oriental, the grand mountain chain that rises from the humid lowlands hugging the Gulf of Mexico. In the higher elevations, where the air quickly cooled, Rolland celebrated the "eternal spring" of towns, including Xalapa.[13] Xalapa was situated well, he thought. It was close and well connected to the more humid but important port of Veracruz, and it was only 186 miles east of Mexico City.

About twenty thousand people called Xalapa home.[14] In June 1925 Rolland set up a temporary residence and downtown office. His children visited on occasion, but they remained in school in Mexico City, likely living with one of the sisters of their recently deceased mother.[15] They would have found that Xalapa residents enjoyed a temperate climate, and that the city was filled with gardens, had a mix of deciduous trees and conifers, and was a labyrinth of narrow, winding streets. Blackberries and hummingbirds thrived.[16] Local teenage girls with braided hair sometimes paid tribute to the Virgin of Mercy at a local cathedral in hopes of obtaining a marriage proposal from their beloved.[17]

The massive Pico de Orizaba volcano towered among the mountains to the southwest. The Cofre de Perote, an ancient volcano, stood closer, just to the west. Sometimes the clouds would linger in the valleys dividing the mountains, which looked black against the oranges and yellows of dawn.

Veracruz was picturesque, but its people had witnessed their fair share of turmoil. American military men had invaded and occupied the port of Veracruz in 1914. Almost immediately after the U.S. withdrawal, the Constitutionalists had established the same port as their base of operations, where Rolland debated land policies and Carranza made his famous revolutionary decrees in early 1915. There had been significant fighting throughout the state. During his first term as governor, Tejada had embarked on an ambitious program of social, housing, and agrarian reforms. Rolland worked on occasion in the state on behalf of the National Agrarian Commission. Tejada and his supporters rallied large groups of unionized allies in the face of often violent opposition from more conservative elements of Veracruz society. Veracruz had also been an epicenter of violence during the De la Huerta Rebellion. Tejada had mobilized agrarian groups to fight against the insurrectionists. Jara's prominent role in defeating the rebellion in the state helped secure his popularity and his rise to gubernatorial office.[18] By 1925, when Jara invited Rolland to Xalapa, the city, and the state it belonged to, had seen widespread political mobilizations.

Utopian Visions

Jara hired Rolland to reimagine Xalapa. The stadium was designed to be the first component of a larger transformation. Rolland and Jara envisioned a scholarly city built around a university, a garden city for the workers, and the stadium, which would unite the neighborhoods around a cross-class public space. It would be a close-knit community of students, educators, and laborers living in collaboration and without exploitation. There would be low-priced housing; communally oriented planning; scientific laboratories; a chlorinated, Olympic-sized swimming pool open for community water exercises; a gymnasium; classrooms; and sports fields. There would be beautiful parks and gardens, as well as a modern electronic communications network connecting the main city, the university, the stadium, and the garden city. The community would be "new, clean, happy, and comfortable."[19]

The project was Rolland's first real chance to bring to life his vision for an ideal community. It echoed many of the ideas that he had laid out in the early 1920s about municipal planning, ideas largely borrowed from American progressives who had in turn been influenced by people and events in Europe and Australasia. Texts like Horatio Pollock and William S. Morgan's *Modern Cities* (1913) emphasized German city management and the new "garden cities" of England. They extolled a new emphasis on collaboration over competition, the "desire for outdoor life," and the "demand for beauty." And they argued that in "civilized" parts of the world "cleanliness, beauty and health have taken up their abode in the modern city and are fast displacing the dirt, ugliness and disease so prevalent in its predecessor."[20] Rolland admired these accounts of clean streets, healthful and thrifty people, and the disappearance of crime and disorder. And although World War I had shaken the certainty of some of his European and U.S. counterparts, Rolland had not lost faith in his belief that through science and effort the world would become more urban and thus healthier, happier, and more prosperous.

The concept of a workers' garden city grew from the progressive experiences shared by Rolland and Jara.[21] The idea was originally developed in the nineteenth century by Ebenezer Howard, an Englishman who had immigrated to the United States, failed as a homesteader in Nebraska, lived in Chicago, and then moved back to England, where he attempted to develop collaborative communities outside of the congested London slums that so many workers called home. Influenced by the utopian ideals of Edward Bellamy's *Looking Backward* and the political-economic theories of none other than Henry George, Howard wrote an influential book titled *To-morrow: A Peaceful Path to Real Reform*

(1898), retitled four years later as *Garden Cities of To-morrow*. His garden cities were to be, as the historian Peter Hall puts it, "the vehicles for a progressive reconstruction of capitalist society into an infinity of co-operative commonwealths."[22] These communities would be connected to larger cities through modern transportation but would be surrounded by green spaces and healthier living conditions. The residents would pool resources, buy cheap agricultural land, and then entice industry to come to them. Optimally they would consist of skilled and unskilled workers: engineers, bricklayers, artists, farmers, architects, surveyors, and manufacturers. The communities would be examples of communal self-governance. In Howard's words, "the citizens would pay a modest rate-rent for their houses or factories or farms, sufficient to repay the capital, and then—progressively, as the money was paid back—to provide abundant funds for the creation of a local welfare state, all without the need for local or central taxation, and directly responsible to the local citizens."[23] It sounds utopian and somewhat anarchist, but it was not a pipe dream. There was an attempt well under way in Letchworth, England, only thirty-four miles from London, to make Howard's ideas a reality.[24] The garden city movement had also influenced urban progressives across Europe and in the United States.[25]

The Xalapa garden city was to be placed just "to the south of the stadium in the admirably arranged hills and valleys." By "practical and scientific means" city planners would resolve the "housing problem" and elevate workers' lives. Rolland's sketch of the community shows a system of curving streets feeding rows of homes, a central administrative complex, a hotel, and a market. The community would be based on cooperation and volunteerism. In Rolland's words, "land speculation based on artificial land valuation would be strictly prohibited. No individual will take advantage of unearned gains but [instead gain through] common effort." The state government would buy the land, and the rent would be based on the cost of basic services and a small fee to the state treasury, which would invest that sum and "intelligently" spend the proceeds on behalf of the workers.[26]

What most people made of this, one can only speculate. Surely many laborers welcomed employment and probably the idea of living in a neighborhood with nicer homes. Residents were familiar with revolutionary pronouncements about equity, cooperation, collaboration, and communal spirit, but few residents would have known anything in detail about the ideas of Ebenezer Howard and Henry George or, for that matter, what the planners intended by building a garden city. This reality touches on an issue that would continue to resurface and hamper Rolland's designs for his entire career: he was not particularly good at working with the people he wanted to help. He talked of improving the lives of all Mexicans, but he was easily frustrated with people who did not live up to his standards or who failed to understand his vision. As a result, he commonly forced his designs on people who often did not understand or agree with them. Many people resented him for this.

Of course Rolland expressed only positives about his plans for Xalapa. He characterized the future stadium as the glue that would bind together the envisioned community. It was intended to bring the various classes and sectors of society together in order to celebrate their common mission of building a better city and a better Mexico. In Rolland's mind people would join together in singing the national anthem and celebrating the athletic achievements of a reinvigorated youth while letting go of the divisions the revolution had produced. His plans for the stadium exhibit his continued movement away from radical politics toward a more conciliatory approach. In order for the country to move beyond violence, conflict, and destruction, the classes had to find a way to build common purpose. Rolland believed this conciliation could be shaped by a built environment, especially through more equitable living conditions and public spaces that promoted health and shared pride.[27]

The Stadium

The stadium, like other components of the larger project, was also influenced by international trends. Stadiums had been around since before the ancient Greeks and Romans, to whom Rolland paid homage with his vision. A number of precontact indigenous cities had ball courts. But it was the late nineteenth century and early twentieth century that ushered in a new era of global stadium construction. These buildings were influenced by the rise of new materials, including steel and reinforced concrete. Athletes from fourteen nations participated in the rebirth of the Olympic Games in 1896.[28] Greece hosted the games in Athens, and government officials ordered the construction of a number of new facilities in preparation for the event. For the subsequent Olympics in Stockholm (1912) and Paris (1924) engineers and laborers designed and built new stadiums. Mexico participated in the Paris Olympics, gaining widespread recognition across Latin America. Although none of the Mexican athletes won medals, many Mexicans saw their participation as a sign that Mexico was one of the more developed nations in Latin America, representing the region on the global stage.[29] U.S. architects and workers meanwhile were putting the finishing touches on the Los Angeles Memorial Coliseum (1923) and Chicago's Soldier Field (1924).[30]

In 1924 Mexican leaders began their own campaign to build large stadiums and sports complexes, simultaneously expanding global trends and Mexican nationalism. Like their more famous Olympic counterparts, Mexican stadiums would serve not only as sports arenas but as a form of mass media and spectacle. They became the stages of massive rallies and populist politics. Stadiums were symbols of pride built with the intention of creating unity, celebrating youth culture, and providing distracting entertainment.

The most attention-grabbing project in Mexico had been the Estadio Nacional, or National Stadium, in Mexico City. It was a massive and expensive project headed by José Vasconcelos, the prominent intellectual whom Álvaro Obregón had made secretary of public education. Vasconcelos commissioned the architect José Villagrán García to design the stadium. Unlike Rolland, Vasconcelos detested modern architectural styles celebrating concrete. He wanted the stadium crafted from marble and stone. This particular aspiration was dashed when the government refused to fund such a costly endeavor. Vasconcelos and the architect subsequently battled over the design and materials, and the stadium ended up being built largely of cast iron and concrete in a "mish mash of styles."[31] The famed artist Diego Rivera provided some of the artwork. Inaugurated on May 5, the anniversary of the famous Battle of Puebla in 1862, when a Mexican army had temporarily defeated French invaders, the stadium received celebratory accolades from artists and writers, including the Stridentists, across the country. It was a symbol of Mexico's new revolutionary order. It became the center of presidential displays, including Calles's presidential inauguration. Unlike the Xalapa Stadium, however, the National Stadium was short-lived. Within months of its completion the structure began to show cracks, which Villagrán García and Vasconcelos blamed on each other. In 1950 Pres. Miguel Alemán (1946–52) ordered the demolition of the stadium.[32]

The origins of the Xalapa Stadium predate Jara and Rolland's involvement. In 1898, only two years after the Athens Olympics, William K. Boone, an American immigrant who had become the president of Xalapa's chamber of commerce, proposed building a stadium. Tejada had worked with Boone and members of the Jalapa Railroad and Power Company to follow through with the idea. The result was a small earthwork stadium and sports area, completed in 1922. It was nothing fancy, but the Xalapa Stadium that Rolland constructed was built on this foundation.[33]

Rolland paid homage to the ancient Greeks and Romans, but he also designed Xalapa Stadium to showcase the new possibilities provided by reinforced concrete. Unlike Vasconcelos, Jara does not appear to have fought with Rolland over the plans for the stadium, which was to stand out aesthetically but also to blend into the local geography. The massive horseshoe-shaped, cantilevered roof would be modern and constructed as one solid piece of reinforced concrete. As Federico Sánchez Fogarty, one of Mexico's most vocal proponents of concrete architecture, stated in the journal *Cemento*, the stadium would be "akin to a colossal stone of great solidity, which could have been sculpted by a mythological artificer."[34] Opposite the seating and U-shaped roof stood Doric columns topped with bronze gladiatorial figures. Other statues of athletes and warriors in neoclassical form surrounded the stadium.[35] Along with the stadium and its artistic embellishments, Rolland built water tanks and a drainage system of reinforced-concrete tunnels, which he wanted to connect to a larger project that would drain the *aguas negras* (runoff and wastewater) of the "ancient city" of Xalapa. He also included a modern lighting system consisting of 153 500-watt lights, so all classes of people could celebrate events day and night.[36]

Construction began on June 28, 1925. Rather impressively, Rolland and approximately six hundred white-shirted laborers—carpenters, metal workers, bricklayers, work supervisors, engineers, and designers—built the stadium in only seventy-seven days, an astounding accomplishment. Along with the setup of rafters, frames, and molds, Rolland ordered the creation of a concrete fabrication plant on the site.[37] Hernández Palacios, then a student at the local prep school, recalled that "laborers worked feverishly, and the mechanical shovels took advantage of the natural basin to shape the majestic stadium."[38] According to another account, laborers worked day and night, sculpting the landscape and making the rigorous preparations needed before pouring the nearly ten thousand square feet of concrete necessary for the structure that would comprise the stands, roof, and connected columns.[39]

The stadium, however, did not come cheap. Although the Jara administration and the press wrote that the total cost was 350,000 pesos, the total cost, including labor and the inaugural ceremonies, was more than 500,000 pesos. The roof alone cost 192,000 pesos.[40] In order to fund the stadium and other projects Jara turned to levying new taxes and to forcing foreign oil companies in the state to pay royalties, a requirement they did not always fulfill. Jara's actions caused tension between a number of oil companies and the Calles administration, further stressing the relationship between the president and the governor.[41] As the town prepared for the opening ceremonies on September 20, a large number of vendors, laborers, and state employees protested because they were still owed back wages.[42] In July 1926, nearly a year after construction started, Maples Arce was still attempting to come to terms with workers upset about the lack of administrative efficiency in paying workers.[43] The state government likewise struggled to pay teachers and bureaucrats.[44] Stadiums were nice, but not paying workers made class conciliation a difficult process.

Praise

The overall turnout for the stadium's inauguration was nevertheless tremendous. The streets of Xalapa buzzed with anticipation, not only for the new stadium but also the arrival of the nation's most prominent political leaders. Despite differences with Jara and Rolland, President Calles kicked off the inauguration. In addition to thousands of locals there were members of the Ministry of War and Marine, Ministry of Foreign Relations, Ministry of the Interior, and Ministry of Public Education, as well as a number of generals in attendance. Diplomatic representatives from China, Switzerland, the Netherlands, Spain, France, Peru, Japan, Chile, the Soviet Union, Colombia, Czechoslovakia, and Belgium were also in attendance. Calles drove out onto the field in a car with María Luisa Apapcaio, the "Queen of the Patriotic Celebrations." Speaking to the audience of more than twenty thousand spectators, Calles congratulated the stadium's creators and the city on the grand achievement while uncovering the inaugural plaque to the cheers of the roaring crowd.[45] Rolland discussed his design and his vision of Xalapa's even brighter future. He praised nature, science, and education: "nature and science provided this temple. [With it, we must] cultivate the spirit and mind because only the educated races harmoniously guide their people."[46] Governor Jara, reveling in the attention, praised Xalapa's new stadium as the foundation of progress.[47]

Much of the display put on during the inauguration focused on the youth, especially their military training and athleticism. The Jara administration had ordered teenagers to receive military training at the local schools. According to one of the event's participants, he and his fellow classmates were proud but not exactly in sync. They dressed in olive green trench coats and carried rifles on their shoulders. Filled with anxiety and overwhelmed by the cheers of the crowd, they struggled to keep time as they marched their way around the track to receive praise from Governor Jara and President Calles.[48] There were also flag exercises, a relay race, and a game of *balón*, in which teams pushed against a giant ball, attempting to force it toward their opponent's goal.[49]

The journal *Horizonte*, which the Stridentists founded the following April, regularly celebrated the stadium and the material with which it was constructed. Indeed the magazine celebrated many of the concepts Rolland held dear. Much of the May 1926 issue was dedicated to the ideas of Henry George. It provided a translation of an essay by George himself and another by his most famous promoter, Leo Tolstoy.[50] Other issues of the magazine showed the stadium, along with Jara's radio station then under construction, as a symbol of progress—the symbol of Stridentopolis. Maples Arce called the stadium "audacious" and the "the most beautiful . . . center for the youth and their formation of intelligence, beauty, and humanity."[51] He also praised the "resistance, durability, and economy" of concrete more generally.[52] Adulation and emotion also radiated from the prominent Stridentist writer Germán List Arzubide, who edited *Horizonte*. In his celebratory piece "Construid un estadio" (Build a stadium), List Arzubide applauded the stadium's ability to unify people and to improve the lives of all Xalapa residents:

> Lift up a stadium, as you would lift an altar for a better life and greater fruits of good and strong men.
>
> Lift the students, so they can go there in strength to renew their souls mutilated by the parasitic life of tortuous schools.

> Lift up the workers, so they can receive distraction that brightens and teaches, liberating them from the bondage of vice.
>
> Lift up the employees, again strengthened by the combat and contemplation that inflames, animating them to undertake a freer and more dignified life.
>
> Lift up the teachers, with the desire for a better youth, more optimistic and more noble.
>
> Lift up the soldiers, with a promise that murder and crime are of the past.
>
> Lift up the women, for there they will prepare for the future struggle in which they will also be workers; so that your children will receive the strength and nourishment of a more intense life.
>
> Lift up everyone, offer your arms and your yearnings, your money and your minds; all of you who prepare your lives not wanting to be mutilated by the lack of liberty to offer.
>
> Teachers, students, workers, employees, those that go through life like shadows in a great night defeated in a world that drowns hopes, searching futilely for a better way, build a stadium.[53]

"Construid un estadio" reflected Rolland's own notions for the stadium's purpose. Through technology, shared material development, and enlightened city planning, people would unite across classes and education levels and build a better and more unified society despite a conflicted past. As one historian later wrote, "The stadium built in 1925 was the theatre par excellence for this symbolic conciliation."[54] On this occasion Rolland seems to have largely succeeded.

The rest of Rolland's vision of Xalapa proved less successful. The unstable politics of the era combined with the high cost of the stadium and the numerous other public works projects under way undermined the grander goals of the Jara administration. The state government ran up deficits, creating a number of detractors at the local, national, and international levels. As a result, progress on the envisioned university, the Universidad

Veracruzana, was slow going. The administration did start on the project before the end of 1925, but it was eighteen years before it resembled a functional university. Less successful still was the garden city, which never came to be. It seems the plan was abandoned because there was no funding to turn the idea into reality. If any ambitions remained to build it, they were dashed in 1927. In the midst of the outbreak of the Cristero Revolt and the political revolt of Francisco Serrano and Arnulfo Gómez, which rattled the state during Obregón's divisive run for a second presidential term, Jara's opponents in the state legislature ousted his government in late September.[55] Jara found no support from Calles. Maples Arce fled after his life was threatened.[56] Rolland, proud of his stadium but realizing that he would not be able to implement more of his vision, had already moved on, deciding to focus on private construction contracts elsewhere and rebuilding his family life.

Rolland's work in Xalapa, more than any other project, placed him squarely within the world of utopian visionaries, modernist planners, and even avant-garde poets. Rolland had been a proponent of concrete architecture for nearly two decades, but the Xalapa Stadium was the first project in which he was able to truly showcase his reinforced-concrete design aesthetic outside of home building and water projects. Influenced in part by classical elements so often infused in stadium construction, Rolland's concrete roof structure was celebrated by Mexican avant-garde artists and modernist architects for its bold and powerful design. Mirroring other engineers and architects of the time, Rolland envisioned the stadium as a prominent piece within a larger vision of city planning: the stadium would be a place that brought people of different classes and social backgrounds together—the centerpiece of a new combined scholarly city and garden city based on education, fairness, cleanliness, physical fitness, efficiency, and interconnectedness.

Rolland's plans for the garden city of Xalapa show how he meshed different intellectual influences that straddled the juncture of the late nineteenth and early twentieth centuries. Motivated by modernist trends under way in Europe and the United States, Rolland had been influenced even more by the utopian and progressive visions of the late nineteenth century. Ebenezer Howard and his U.S. admirers and their attempt to build cooperative commonwealths based on pooled resources, new technologies, and simplified taxes in new suburban settings directly shaped Rolland's vision for Xalapa. Howard himself was building on even earlier visionaries. He had been preceded by French utopians such as Henri de Saint-Simon and the nineteenth-century Russian writer Nikolai Chernyshevsky, who pushed Russians to become a "new people" through embracing modernity in more rural and suburban settings.[57] Another significant influence on Howard had been Henry George. That influence was something Howard and Rolland shared, and it surely influenced Rolland's embrace of garden-city planning.

Rolland's work in Xalapa thus shows an interesting combination of progressive, utopian, and modernist designs. Rolland's stadium, praised in Mexico's modern architectural magazines, fit well within the modernist world of his contemporaries in other countries, people including Le Corbusier and Frank Lloyd Wright but also the more classically influenced John and Donald Parkinson, the father-and-son architectural team that designed the Los Angeles Coliseum. Rolland's ideas for urban planning, however, borrowed extensively from late nineteenth-century planners who were interested in lessening the social negatives of urbanization but also in creating modern industrial nations out of "underdeveloped" countries. Rolland knit together these worlds, borrowing from his own late nineteenth-century education, his experience with U.S. progressives, and his interaction with modern engineering, architectural, and stadium-building trends. He put them together in a fashion that was comfortable to him, but in a way that he felt best fit the context of Mexico. To Rolland, Mexico was a country in the infancy of its development into a more urban, industrial, and modern society—a country where intellectual leaders were striving to build a modern and unified people out of revolutionary factionalism and a multitude of different cultures and languages.

Only recently ousted from the national government in Mexico City, Rolland had an ability to pivot toward Xalapa, design a garden and scholarly city, and build a stadium with astonishing proficiency that was impressive even if the outcome fell short of its original grand conception. The stadium showed what Rolland was capable of achieving when a government supported his projects and did not interfere with his work. The failure of the garden city showed the limitations Rolland continued to face. Jara had difficulties paying the workers building the stadium, the very workers meant to benefit from the future garden city, which ultimately the Veracruz government could not afford. In addition to problems with money, the aftershocks of the revolution failed to subside. The assassination of Obregón and continued revolts rocked Mexico, ultimately leading to the ouster of Jara's government. Rolland left shortly after the stadium's completion, as he realized that he would be able to accomplish little else. But he would remain enamored with the Veracruz highlands, holding onto visions of a better life in the land of eternal spring.

9

Mr. Bothersome

CONSTRAINTS HAVE A WAY of forcing people to reevaluate their position in life. In 1926 Modesto Rolland found himself back in Mexico City. He had completed the Xalapa Stadium, but it had become apparent that Gov. Heriberto Jara would not be able to afford the other large-scale projects Rolland had envisioned. Rolland's dream of developing the ideal city was, for the time being, dashed. In the nation's capital Rolland, with some hired caretakers, focused on getting his children through school, caring for his sister, and working in the private sector. He was forty-five and had gained a little weight, and, contrasting sharply with the rest of his thick black hair, large streaks of white began to dominate much of the left side of his head.

As the next decade dawned Rolland became louder about his frustration with the path of the revolution. He found the populist politics of Álvaro Obregón and Plutarco Elías Calles to be little more than pandering by politicians who placed political expediency ahead of long-term planning. Rolland still carried a grudge about the closing of the free ports. He found the government's continued reliance on tariffs and indirect tax policies unwise and, combined with short-cited agrarian reforms, damaging in an already problematic global economic environment. He also argued that clientelist politics had led to a bloated and ineffectual bureaucracy.

Like his city planning and architectural designs in Xalapa, Rolland's words reveal how his past meshed with more contemporary issues to influence his worldview. In some ways, such as his unyielding dedication to the ideas of Henry George, Rolland was somewhat unique. In other ways, such as his disdain for personalist politics and political extremism, he was similar to other moderate technocrats of the revolution, those people who had established their reputations as pragmatic scientists, developers, and economic planners but who felt themselves unheard in a world of increasing protests, strikes, and populist politics.[1]

Excitement and Turmoil

Mexico City had grown into a vibrant urban center. There were more cars, more electric signs, and more buildings. People from around the world had long visited Mexico City, but there was a renewed sense of metropolitanism. *Flapperistas* abounded, jazz mixed with *corridos*, and movie theaters played the best foreign and domestic films. A host of foreign intellectuals had flocked to the capital to witness and participate in the revolution. Mexican artists and writers flourished.[2] Mexico City had become a bohemian mecca of sorts.[3]

During Calles's first year in office the economy did relatively well. The Calles administration established the Banco de México, which became something similar to a central bank. It was established with the help of the lawyer-turned-economist Manuel Gómez Morín, who had studied the central banking system in the United States while on a diplomatic mission for the Adolfo de la Huerta administration.[4] Calles had an engineer, Rolland's fellow technocrat and slightly older classmate Alberto J. Pani, serving as head of the Ministry of Finance. Pani, like Rolland, had been influenced by progressive thought, though he possessed a more traditionally liberal economic worldview.[5] Industrial output had largely returned to prerevolution levels. The production of textiles, beer, steel, and cement was increasing.[6]

The rosy outlook did not last long. In 1926 the economy took a downward turn that continued until the early 1930s. Foreign oil interests started to invest less and leave. They were increasingly perturbed by Mexican policies and instability, but these business operators were even more compelled by the exhausting of easy-to-tap oil and by new opportunities in Venezuela. As Europe recovered from World War I, prices for agricultural exports dropped. In 1929 the U.S. stock market crashed, throwing much of the global economy into a depression. All the while working-class organizations rose in greater numbers and expressed more discontent, which Calles simultaneously attempted to curtail and to use to his political advantage.[7]

A number of revolts—aftershocks of the revolution—also continued to plague Mexico. In 1926 the Calles administration, affiliated unions and agrarian groups, and a number of state governors battled the Catholic Church and organizations affiliated with it over the government's attempt to enforce greater restrictions on public displays of Catholicism and to break down Catholic resistance to secularization programs. Some among the faithful in Mexico City saw an ominous warning in a local cathedral, where there was a cross that they believed God caused to shake.[8] This backlash against a broad swath of revolutionary policies, which was especially active in northwestern and central Mexico, became known as the Cristero Revolt.[9]

During the run-up to the 1928 presidential election Álvaro Obregón influenced his supporters in the Congress to tweak the Constitution to allow him a second, nonconsecutive presidential term, providing him another opportunity to run for high office. He subsequently won the election, but he never again sat as president. On July 18, 1928, while celebrating his victory with colleagues at a Mexico City restaurant, Obregón allowed an artist named José de León Toral to draw his portrait. Instead of getting his sketch, the president-elect received five bullets. Slumping from his chair, Obregón died immediately. In addition to being an artist de León Toral was a young religious fanatic who believed he was on a mission from God to end government persecution of Catholics. His trial became a massive spectacle, but even that failed to calm tensions, as Calles had hoped it would.[10]

Knowing that Mexico once again teetered on civil war, Calles handled the situation with considerable skill. In that year's annual presidential address to the Congress, Calles lamented Obregón's death, but he used the situation to call for an end to *caudillo*, or strongman, rule. Calles assured the legislature that they would build a new regime based on "institutions and laws."[11]

Calles also gathered the top military leaders together and proposed that they, along with civilian leaders, work together to hammer out differences without reigniting the civil wars that had devastated the country. Calles astutely supported the nomination of Emilio Portes Gil as interim president. A lawyer and former governor of the northeastern state of Tamaulipas, Portes Gil was not considered a close ally of Calles or Obregón, thus allowing the camps that had developed around each figure to come to a compromise.[12] A new election would be held in 1930 after things had stabilized. In the meantime Calles promoted the creation of a truly national party in which disputes could be worked out internally and relatively peacefully. The resulting Partido Nacional Revolucionario (PNR), or National Revolutionary Party, would become the foundation of the single-party state that would dominate Mexico for the next seventy years.[13]

Portes Gil's presidency, however, did not satisfy all military leaders. Some generals who were close to Obregón harbored suspicions that Calles was behind the assassination.[14] A number of military officials were also upset at Calles's move to place civilians into the presidency. Calles backed the engineer and ambassador Pascual Ortiz Rubio as the PNR candidate for the 1930 election. On March 3, 1929, Gen. José Gonzalo Escobar and 28 percent of the armed forces rose in rebellion. Escobar claimed that Calles, who still possessed immense influence, was essentially placing Ortiz Rubio in the presidency. The rebellion was significant, but it lacked clear goals outside of toppling the government. Calles, serving as secretary of war, personally led the government response, quickly putting down the insurrection. It was a huge blow to would-be rebel generals, but it prolonged the ongoing Cristero Revolt, which Portes Gil, Calles, and other top political leaders realized needed to end in order to assure stability. The Portes Gil administration, papal representatives, and local officials of the Catholic Church came to terms shortly thereafter with the help of the American ambassador, Dwight Morrow.[15]

In 1930 and early 1931 the Congress debated legislation that significantly changed the political organization of Baja California. The peninsula was still sparsely populated as a whole, but the population in the north of the territory was expanding alongside tourism, mining, and irrigation activities. The Congress split the Territory of Baja California into two territories, dividing the peninsula along the lines of the two districts that already made up the territory.[16]

Throughout this excitement and turmoil Rolland spent his time putting his own house in order, eventually returning to the public sphere as events calmed. Yet he still directed some of his private work toward what he saw as national priorities. He invented a wind-powered water pump that he hoped farmers could use to irrigate lands where there was little access to electricity. During the 1930s he involved himself again in railway projects that would connect southern Mexico to the Tehuantepec Railway. He eventually obtained high-ranking positions in the Ministry of Communications and Public Works. Again he pushed the free ports concept while becoming even more critical of revolutionary politics. This frustration rubbed many of his acquaintances, employees, and employers the wrong way. Playing on his name, they sometimes referred to him as "Don Molesto," or Mr. Bothersome.[17] As his children grew into adulthood and his annoyance with politics increased, he became more concerned with their personal advancement and his own material success. And although he never became as crooked or as rich as some of his peers, this shift led him onto a path not always in tune with his earlier idealism and his critiques of crony capitalism.

The Piano Teacher

At some point in 1926 Rolland thought it would be a good idea for his daughters, Enriqueta and Martha, to take piano lessons. Enriqueta was going on eighteen and Martha was fifteen. Rolland enjoyed music, classical music mostly, but also popular songs. Playing piano was also something that showed a certain level of class and education. It was a skill that not only brought enjoyment but also one that many women from well-to-do families mastered to some degree. Perhaps he thought his daughters needed a hobby. Or perhaps he wanted to get closer to their piano teacher.

The woman Rolland hired to teach his girls was Rosario Tolentino Morales. Tolentino was young, in her twenties. Her hair was bobbed in the fashion popular with so many urban women of the decade. She came from an upper-middle class family that had lost much of its fortune during the revolution. Tolentino had dreams of becoming a concert pianist. She majored in music at college, and the only thing preventing her completion of a degree was a required recital with backing musicians. Since she did not have the money to rent a concert hall or to pay orchestral members, Rolland agreed to pay for them. Tolentino and Rolland were married the following year.[18]

Rolland appears to have been happy with Tolentino, but the children, according to at least one of Rolland's descendants, initially cared little for the marriage. Rolland and Tolentino traveled frequently; the children "suffered." When Rolland and Tolentino were away, the children sometimes stayed with one of Garza de Rolland's sisters.[19] Photos from the time, however, show no signs of abuse. The children appeared healthy, sharply dressed, and full of privilege. Soon Rolland and Tolentino would bring a new addition to the family, a girl named Ana María.

The Private Sector

To pay for all those travels Rolland took on a number of projects. His emphasis on nongovernment work was of course driven by other factors as well. In addition to building and remodeling a number of homes, Rolland constructed a building just outside the Mexico City limits for the Foreign Club, which served as a gambling hall. The Foreign Club was a sister casino to another club, located in Ensenada, Baja California. According to his descendants, Rolland was admittedly against gambling. It went against his proclivity toward order, control, and hard work. It appears that at this point in his career, however, Rolland became willing to put aside at least some of his moral qualms in order to make a comfortable living.[20]

Rolland designed and oversaw the construction of another large building project in the 1920s—the Hotel Chula Vista in Cuernavaca, Morelos. Fifty miles south of Mexico City, the white hotel stood out against the backdrop of mountains. The sprawling five-story building looked like a giant hacienda, with Spanish roof tiles and massive arched doorways and windows. There was a restaurant with an outdoor patio dining area and a rectangular swimming pool with a diving board. The Chula Vista became a favorite getaway spot for wealthy foreigners and Mexico City residents.[21]

In 1930 Rolland made personal investments in the highlands of Veracruz, buying the Minería del Rosario, a mine in the Las Minas region of central Veracruz, about thirty-five miles northeast of Xalapa. Perhaps he thought that, since the government had failed to develop Mexico in a beneficial manner, he could do the job himself. He surely saw a potential for profit. The area had significant deposits of copper and gold. Rolland would later use the mine to help create synthetic fertilizers. He had not fallen out of love with the region's natural splendor; he enjoyed being there in the narrow and mist-filled valleys.[22]

While in Mexico City, Rolland tinkered with an invention he had long been working on—the Aero-Motor México. The product was a wind-powered water pump, which he advertised as "an entirely new way to harness the power of air efficiently."[23] According to the 1932 pamphlet Rolland used to market the device, it consisted "of a metal tower that rotates on ball bearings at the bottom and is held in position at top by another bearing that is connected to six galvanized, steel cables that anchor firmly to the ground. The drive wheel comprises three blades arranged scientifically in cylindrical form and suspended by two galvanized steel cables, running from inside the tower, connected to a winch that exists in the base. In this way, the wheel can slide along the pole or tower to its working position to the ground."²⁴ Rolland argued that his specially manufactured pump, with two pistons, was superior to the pumps in the United States, which he claimed were less efficient. Indeed Rolland claimed his device pumped more than three times the amount of water other equipment did. His Aero-Motor pump also cost less than the average American model. According to Rolland, the average sixteen-foot wind-powered water pump cost $2,400. His pumps, on the other hand, started at just $1,500, though upgraded versions cost a bit more.²⁵

Rolland thought that efficient water pumps were crucial to Mexico's progress. Influenced by his childhood in the arid climate of the Baja California peninsula, Rolland imagined that if farmers could access underground water efficiently then the desert regions of Mexico could bloom, much like irrigation projects had turned California into the fruit basket of the United States. His device, powered by wind, would also help farmers who had limited or no access to electricity. Most of Mexico's main urban centers were electrified, but rural villages often had no consistent access to electric power.²⁶

Rolland wrote to Calles and to media outlets hoping that they would see the potential in his invention and help advertise it. He told them that the final result was the product of twenty years of thought and work.²⁷ He had definitely been playing with the idea for a long time. Thirteen years earlier Rolland had registered his first patent for a wind-powered motor.²⁸ Writers for *El Nacional*, the official newspaper of the PNR, promoted the invention as the "most immediate and practical solution" to irrigating small and medium-sized farms. According to the newspaper, Rolland had arranged for his own public exhibition of the pump, too. He set up a model outside a small workshop on the corner of Baja California Street and Calzada de la Piedad Avenue. People stared at the "iron tree" with "astonishment at the spectacle of the strange cylinder that spun without cessation."²⁹ Surely the spinning air-powered pump set up on a city street corner, looking very much unlike other windmills, caught the attention of passing visitors and urbanites, but few of them bought one. There are no ads or discussion of the machine in Rolland's own accounts after 1932.

Discontent Made Public

In 1930 the Henry George Foundation of America had invited Rolland to speak at their annual conference in San Francisco, California. The fog and trolleys moved up and down the hills overlooking the bay. Ocean trade had become immense, and the streets were filled with people speaking diverse languages. The wharf and downtown market were vibrant, smelling of fish, saltwater, eucalyptus, and gas-powered engines.

At the convention Rolland gave his take on what he saw as the bitter fruits of the revolution and the difficulties he faced in establishing a Georgist republic in Mexico. He discussed the abuses and downfall of the Porfirio Díaz administration and provided a brief outline of his work with Carranza and Obregón. Mostly, however, Rolland bemoaned what he saw as the failure of Mexico's *ejido*, tax, and industrialization policies.

After President Obregón forced him out of the National Agrarian Commission, Rolland had watched with dismay the politicization of the ejido process. Rolland told the audience that there was justice in returning communal lands to Indian communities, but he insisted that the haphazard donation of millions of acres for political gain, without careful economic consideration and scientific planning, was having disastrous results. There was a 500-million-peso and rising "agrarian debt," which federal taxpayers would have to pay, adding to what Rolland saw as "an infinity of taxes." He argued that the debt was not necessary because, if a single-land tax had been implemented, hacienda owners would have freely returned much of their unused land to the government to avoid taxation. Rolland further lamented the decline in agricultural production, which had become so bad that Mexico was importing corn from Africa.[30]

Rolland also condemned the government's ejido policies for damaging Mexico's middle class, the environment, and the potential for lasting peace. Rolland told the gathered Georgists that "many middle-class investors in rural mortgages had been impoverished." He discussed how in Morelos timberland that had previously been "conserved by intelligent farmers" had now fallen into the hands of "reckless peasants, whose only ambition is to quickly harvest the lumber, irrespective of forestry considerations." As a result, Mexico was facing economic disaster, environmental degradation, and "moral anarchy." Rolland believed that the government had been irresponsible, letting a "barbarous population" steal land and reap the benefits of the former owners' harvests. The process, Rolland continued, fueled violence by catering to "the unbridled desires of the greedy and unscrupulous villagers," which caused factions to develop among villagers, who turned on each other in anger or even violence.[31] If Rolland had little faith in the top political leaders, he possessed even less faith in the skills and wisdom of impoverished Indian farmers.

As for industry and taxes, Rolland tied them together. He argued that the federal, state, and municipal governments had expanded their bureaucracies and in turn their budgets. The federal budget was three times what it had been under Díaz, Rolland complained. This was a problem because in order to fund themselves, governments turned to indirect consumption levies, such as sales taxes, while at the same time increasing protective tariffs to insupportable levels. The result, Rolland stressed, was that cotton cost three times the amount that it did under Díaz and that Mexicans spent larger sums of money paying taxes on food, clothing, and shelter. In a resounding statement of his increasingly negative assessment of the revolution, Rolland told the gathered crowd that

> the revolutionary bomb of ejidos for saving the peasants has so far only proved a dud which has aggravated their present impoverishment. Meanwhile, the urban workmen have killed the goose of the golden eggs, since there are no longer any new factories and the existing ones try to flee if they can. Finally, the security of both life and property, outside of

the few policed cities, has been decreasing steadily as a result not only of the aftermath of a long civil war but of the class struggle which both our agrarian and syndicalist politicians have stimulated for their own selfish ends.[32]

It's evident that Rolland had become embittered over his inability to carry out the reforms he had envisioned for Mexico in the 1910s. He was infuriated that political, military, and union leaders had stood in his way. He was angry that so many of the revolution's powerful figures had prioritized their own wealth and self-aggrandizement at the expense of what he saw as wiser, more just policies. He was hurt because the benefits of the revolution that he had earlier promised with sincerity had not materialized. He had grown wary of class warfare, and he began to doubt the revolution.

Rolland was not alone. Others shared similar sentiments. The economists Gonzalo Robles and Daniel Cosío Villegas became, according to the historian Susan Gauss, dismayed at how well-intentioned policy making fell "victim to economic crises, political intrigues, and the uncoordinated development projects of multiple state agencies."[33] Minister of Finance Pani had risen to his position irked by what he saw as political influence on poor economic choices and backroom deals made by de la Huerta, his predecessor. Pani didn't share Rolland's Georgist sympathies, but he did move the Mexican government away from regressive taxes and toward more progressive, direct taxes, including Mexico's first income tax. Regional *caudillos* and business leaders fervently resisted the tax, arguing it was an assault on regional autonomy, tradition, and individualism.[34]

Rolland's 1930 address in San Francisco also sheds more light on Rolland's notions of progress. He painted the development of Mexican society, land policies, and economics in racial and hierarchal terms that harkened to old notions of Indian primitiveness and European civilization. Rolland borrowed heavily from the American anthropologist Lewis H. Morgan, whose *Ancient Society* (1877) argued that there were three stages of progress: savagery, barbarism, and civilization. Rolland supported protecting Indians, but he painted them as "living in the middle stage of barbarism . . . where private property was unknown." Indians were "a primitive race" that made up 40 percent of Mexico's population. He urged his listeners to consider the "more civilized mestizos (half-breeds) and whites" who together made up the other 60 percent. The ejidos provided increased justice and freedom to "Indian serfs," but Rolland contended they hurt mestizos and whites because Indians were foolish and poor agricultural producers, impoverishing Mexico's mestizos and whites, who were "the principal producers and consumers of the nation."[35]

Rolland's brand of progress and racism didn't consider Indians unredeemable. Like Díaz's minister of education, Justo Sierra, as well as Rolland's contemporaries Manuel Gamio and José Vasconcelos, Rolland believed that racial mixing could bring about an improved, modern people, but in the end these people would be less Mixtec or Tarahumara, slightly whiter, and more Mexican. They would embrace Western science and education, vote in national elections, and participate in the global marketplace. The ejido, Rolland thought, was regressive, not progressive. He sympathized with the anger of Indian communities over having their lands stolen, but he saw Mexico's future as mestizo, unified, and not based on ejido agriculture. Despite his harsh critique, however, Rolland still believed that Mexico remained "one of the most propitious fields for the early establishment of a Georgist republic." He ended his speech with a reaffirmation of his Georgist faith: "I salute you, apostles of the international church militant, in the name of our revered apostle. Henry George."[36]

In 1932 he wrote an essay titled "Comunismo o liberalismo?" (Communism or liberalism?). Building on his address at the San Francisco Georgist convention, he expounded further on why, to him, the revolutionary labor and agrarian movements had failed to benefit public welfare. Rolland attacked the rise of collectivism, which he saw as backward and a threat to individualism and democracy. Although Rolland supported the work of revolutionary leaders in limiting exploitation by foreign capitalists, he argued that they had essentially thrown the proverbial baby out with the bathwater. "Good" capitalists and the trust of foreign investors, he contended, were crucial to Mexico's development, and labor reforms had killed all initiative by encouraging disorderly and unrealistic labor demands while discouraging capital investment and industrial growth.

He considered the failure of revolutionary leaders to create genuinely autonomous and democratically functional municipal governments particularly devastating. Politicians had failed to establish proportional elections, the initiative, the referendum, and the repeal; instead they instituted a corrupt tributary system. He likened the revolutionary government to an extortion racket, a "tyrannical" force that abused "the people of their hard work from the cradle to the grave."[37] This clientelist system, Rolland argued, put in power incompetent people who established ignorant agrarian and labor systems that were not scientific and failed to guarantee high productivity or lift workers out of poverty. The result of this poor governance and economic planning had caused widespread disillusion, anger, and unrest, further destabilizing an economy already reeling from the pains of the Great Depression.

The Great Depression, along with other complicating factors stemming from industrialization and World War I, had caused laborers to back undemocratic governments in other parts of the world, particularly in Germany and Italy, where people put in power fascist regimes, and in the Soviet Union, where earlier dissatisfaction had caused people to put in place a communist dictatorship. Rolland feared that government incompetence intertwined with an already difficult economy would fuel communism, which he perceived as contrary to democracy and the public good.

Rolland's critique of communist rhetoric in Mexico had less to do with the theories of Karl Marx and more to do with how opportunistic politicians used communism as a way to take advantage of suffering workers to bolster their own power. By creating a clientelist ruling class, these politicians exacerbated inequality and misery; they did not relieve it. To fight this threat to individual freedoms and democracy, Rolland argued that there had to be a more genuine, "radical, scientific, and rational fight to end misery," a fight based on his long-espoused progressive and Georgist prescriptions.[38]

There was nothing particularly new about his critique—he had promoted a Georgist version of liberalism and complained about ignorance and corruption for some time—but he was bold for criticizing the government so blatantly. And he didn't stop with government officials. He criticized agrarianists for what he saw as their past-oriented collectivist thinking, and he attacked labor leaders for swindling workers and using the strike to their own advantage. There is no doubt that exploitative and clientelist practices continued with zeal. There were also corrupt practices among agrarian and labor leaders. But Rolland's criticism left little room for allies. He attacked just about every sector of Mexico's government and economy.[39]

Rolland's attack on communism, and the direction of the revolutionary government more generally, stemmed from a number of developments. Looming large were the intense, chaotic, and often violent labor and agrarian movements. As the historian James D. Cockcroft has noted, "By the early 1930s, class war was intensifying and shaking the very core of Mexican society and politics."[40] Peasant revolts and worker strikes—fueled by revolutionary promises unfulfilled, clientelism, radical rhetoric, and a struggling economy—were commonplace. Official communist organizations had been relatively disorganized and weak in the 1920s, but they had grown in membership and influence by the early 1930s.[41] One of the last things Portes Gil did as president was to sever ties with the Soviet Union over the attempts of Soviet agents to form better-organized communist cells.[42] Those communists and anarchists who were true to their ideology, which was often not the case, also opposed capitalism itself, and thus business and industrial leaders as well.[43]

A dictatorship of the workers or class warfare held little appeal for Rolland. Eager to establish a functional and stable society capable of limiting violence, developing infrastructure projects, increasing trade, and improving living standards, Rolland believed that cross-class collaboration was better than dragging out revolutionary upheaval. There needed to be a period of national healing and economic and state consolidation in order to move Mexico forward. Improving Mexico was for Rolland a matter of cleaning up the corruption and clientelism of the government in place, not overthrowing the bourgeois order.

Rolland's idealism was losing its luster. He had lost faith in political leaders and much of Mexican society. Peasants and laborers, not surprisingly, had not always been eager to go along with top-down directives that they saw as foreign and controlling. But in Rolland's eyes those working Mexicans who resisted the initiatives he was a part of were ignorant about how to solve the difficult structural issues underpinning Mexico's most entrenched problems. Agrarian and industrial workers demanded land or better pay and working conditions but often squandered what they gained. Instead of becoming productive farmers they chopped down forests and spent their meager profits frivolously. Labor strikes were destructive and slowed the progress that would bring more long-term solutions to the Mexican people. Top political leaders lacked foresight. They had failed to carry out infrastructural projects responsibly, doling out leadership positions based on friendships or closing ventures down altogether before they had a chance to prove their success or failure. When government officials supported peasant and working-class movements, it was with short-term handouts at the expense of tougher, long-term solutions. Instead of fixing problems, top politicos fanned class warfare to benefit their own political agenda. Although there was a lot of truth in his critiques, Rolland never acknowledged his own arrogance, his knack at upsetting colleagues, or his failure to collaborate collegially with the people whose lives would be dramatically changed by his projects.

Frustrated, Rolland drifted toward a more conservative worldview. He had never cared for the violence and disorder that the revolution had unleashed nor had he thought highly of unions. But as a young revolutionary he had been much more idealistic. He had characterized capitalists as evil vampiric octopuses strangling Mexico. He had pushed for greater state regulation on behalf of Mexican workers. He had had faith that Mexicans of all classes and cultures would come together and take advantage of the revolution to construct a happier and more successful Mexico. The Rolland of 1934 found such thinking naïve and, at least in part, incorrect. Rolland had not lost faith in progress, but he came to believe that successful organizers, including powerful capitalists, would be critical to funding and bringing large-scale projects to completion. Gravitating toward the free-trade aspects of Georgism, Rolland doubled down on his drive to establish free-trade zones and to minimize costs and barriers for developers. And since he could rarely count on others, he surmised that he would have to do more himself.

Returning to the Rails

Grievances in hand, Rolland slowly worked his way back into government circles. Calles continued to hold sway over politics from behind the scenes, but political shifts brought Rolland new opportunities. Early in the decade he took a position with the Lineas Férreas de México, or Railway Lines of Mexico, which was a private company that worked closely with the Ministry of Communications and Public Works. He would soon find his way back into a position in that ministry.[44]

Rolland oversaw work being done by the company on the Ferrocarril del Sureste, or Southeastern Railroad. He headed the construction of the line that was to connect the port of Campeche to Puerto México (which would be renamed Coatzacoalcos in 1936) and the Tehuantepec Railway. The goal was, as Rolland had been pushing for many years, to provide a better outlet for the many tropical resources of the Yucatán peninsula and other parts of southern Mexico. Construction of the line was a difficult task. The railroad would have to cross the swamps and marshes of Tabasco in order to connect Yucatán to the center of Mexico. The Railway Lines of Mexico entity, which had also taken over the Tehuantepec Railway, was also underfunded and, according to Rolland, poorly led. The project was still incomplete by the time Rolland took a new government position in 1935.[45]

Rolland's role within the Railway Lines of Mexico allowed him to once again influence infrastructural development, the Tehuantepec ports, and trade. He reentered debates about the "failed" free ports project, which he continued to promote and defend vigorously. He called their premature closure a "great administrative error." The free ports commission, according to Rolland, had done a solid job in revamping the ports and the Tehuantepec Railway in the face of rebellions, protesting workers, and conflicted Americans. Obregón had destroyed the enterprise just as Rolland had attracted interest from international capitalists, shattering one of the few genuine attempts by the national government to build the economy of southern Mexico. Since then the free ports had remained closed and the regular fiscal ports themselves had fallen into serious disrepair. The inner port of Salina Cruz had filled with sand.[46] It was not clear that the free ports would have been a success, but it was clear that the policies during the late 1920s and first years of the 1930s had failed.

In some ways Rolland mirrored the politicians he seemed to get along with the least. He cared little for Calles, but they were similar in their critique of communism, agrarian policies, and the labor movement. Although Calles built populist, corporatist coalitions in order to better secure the government and his own personal rule—something Rolland abhorred—Calles moderated in office, proving to be far from the Bolshevik many U.S. conservatives feared him to be. The same year that Rolland was complaining to the Georgists in San Francisco about the failure of ejidos, Calles had criticized them, too, calling revolutionary agrarian reform a failure and pressuring the government to shift its support to individual (often large-scale), market-oriented farmers. Even more so than the subsequent three presidents he influenced, he cracked down on radical groups, especially communists. Those presidents leaned toward more conservative, liberal economic policies, even while incorporating large unions into their political networks.⁴⁷ Where Rolland was different was that he had developed a serious distaste for the political system in place, especially the backroom dealings of the PNR. Despite the fact that Rolland had become rather politically astute himself, his longtime claim to being apolitical genuinely came from a dislike of politics. Rolland wanted things done. Politics included too much talking, palm greasing, and red tape.

He shared this frustration with a number of technocrats from his generation and later generations. Gómez Morín had become the first person to chair the board of the Banco de México, and in doing so he thought he would end corruption and favoritism between banking and politicians. Instead he saw the bank become a bastion of favoritism and shady loans to political leaders, including Calles and Pani. Jaded, Gómez Morín resigned his position in 1928. He would go on to become one of the leading founders of the Partido Acción Nacional, or National Action Party, in 1939. In his view the party stood opposed to the single-party state and was a pragmatic organization disinterested in leftist and rightist politics, thus being similar to Gaullism in France. However, the party would gain a rightist reputation, largely for its supporters in the Catholic Church and the party's ties to business interests. In 2000 PAN became the first party to defeat the Partido Revolucionario Institucional (PRI, the ultimate successor of the PNR) in a presidential election.⁴⁸

The development of single-party politics that had grown out of the PNR also disturbed Rolland. He remained a fervent supporter of a more transparent and truly democratic government, but he would, like most other revolutionary officials, become a cog within the state the PNR dominated. The presidential election of 1934 brought to power another PNR candidate, Lázaro Cárdenas. It represented a backlash against the conservatism of the three short-term presidents who had preceded him, as well as a renewed leftward political shift. Under Cárdenas's leadership strikes increased to new levels. Cárdenas nationalized oil companies in November 1938, gaining the support of unions and communists and the ire of British and U.S. capitalists. Yet it was Cárdenas who brought Rolland back into prominence as a government official, providing him his highest-ranking government positions and significant influence over Mexico's economic and infrastructural development. As the inclusion of Rolland clearly demonstrates, Cárdenas's politics and designs for Mexico were complex and could even at times appear contradictory.

For his part Rolland continued to walk a path in the middle, attempting to implement what he saw as moderate, modernizing, and apolitical developmental policies. This benefited Rolland at times, while at other times it hurt him. Mixed with his genuine talent and enthusiasm for connecting and building Mexico, Rolland's nonpartisan stance made him compelling to political leaders like Cárdenas, who needed skilled technocrats to help carry out his vision of a more industrial Mexico. At the same time, Rolland's continued critiques of the revolution, of people he saw as pandering political bosses, of imperialist foreign capitalists, of ignorant peasants, and of corrupt union organizers inflated his persona as Mr. Bothersome, making him enemies on the left and enemies on the right.

FIG. 26. The bust of Heriberto Jara outside Xalapa Stadium. Photo by Koffermejia, Wikimedia Commons.

FIG. 27. Construction of Xalapa Stadium, 1925. Photo courtesy of Jorge M. Rolland C.

FIG. 28. A drawing from Modesto Rolland's plans for the Xalapa garden city, 1925. From *Jalapa-Enríquez: Sus obras; la Universidad Veracruzana, el Estadio, la Ciudad Jardín* (Xalapa: Gobierno del Estado Libre y Soberano de Veracruz-Llave, 1925).

FIG. 29. Wedding photo of Rolland and Rosario Tolentino with the Tolentino family, 1926. Photo courtesy of Deanna Catherine Wicks.

FIG. 30. Tolentino, Rolland, and Rolland's children, 1927. Photo courtesy of Jorge M. Rolland C.

FIG. 31. Tolentino and her child with Rolland, Ana María, 1929. Photo courtesy of Deanna Catherine Wicks.

FIG. 32. Hotel Chula Vista, c. 1930. Postcard courtesy of Jorge M. Rolland C.

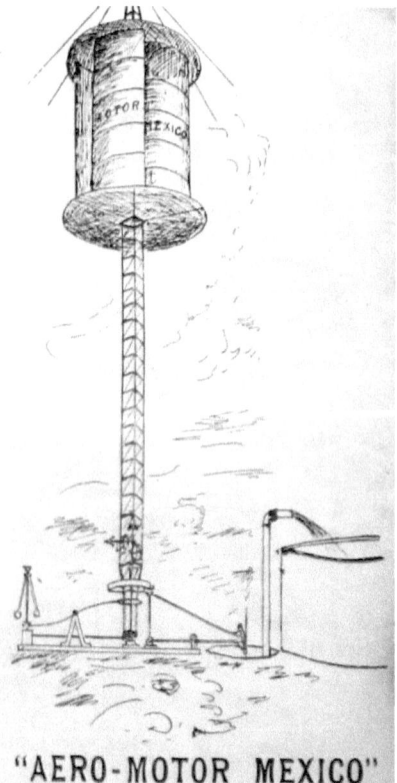

FIG. 33. Pamphlet for Rolland's Aero-Motor México, a wind-powered water pump, 1932. Courtesy of Jorge M. Rolland C.

FIG. 34. Tolentino and Rolland, ca. 1935. Photo courtesy of Deanna Catherine Wicks.

FIG. 35. Tolentino, Ana María, and Rolland in Xochimilco, 1938. Photo courtesy of Deanna Catherine Wicks.

FIG. 36. Illustration of the ship railway envisioned by the U.S. civil engineer James Eads, from an 1880 issue of *Scientific American*.

FIG. 37. (*opposite top*) Illustration of the ship railway envisioned by Rolland, 1946, from Rolland's *Transporte de buques por el Istmo de Tehuantepec*. Used with permission by Jorge M. Rolland C.

FIG. 38. (*opposite bottom*) An imaginative and simplified artist's rendition of the ship railway concept, from "Wants Ship to Take 'Overland Route,'" *Nonpareil* (Council Bluffs IA), April 25, 1949, 12. Photo courtesy of *Omaha World Herald*.

FERROCARRIL DE 10 VIAS, PARA TRANSPORTE DE BUQUES
A TRAVES DEL ISTMO DE TEHUANTEPEC

EL BUQUE HA ENTRADO A LA CAJA FLOTANDO, PROCEDIENDOSE ENTONCES A PONER LAS COMPUERTAS
TANTO DE LA CAJA COMO DE LA ESCLUSA. ENTONCES LA CAJA PUEDE COMENZAR A MOVERSE USANDO
SU POTENCIA PROPIA.

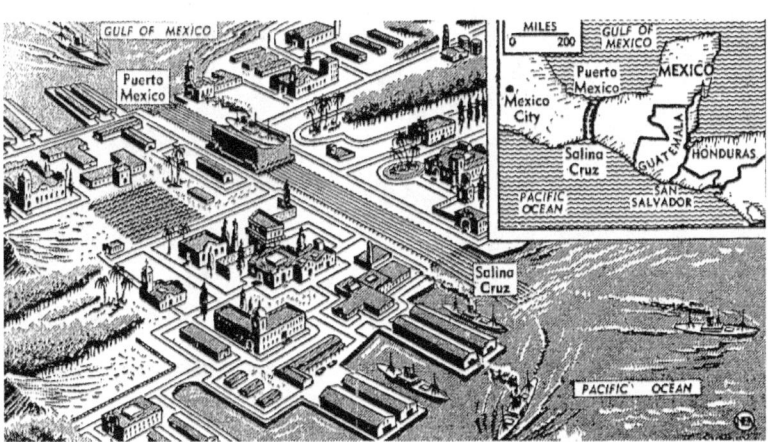

FIG. 39. View of Córdoba, Veracruz, as seen from Rolland and Tolentino's Rancho Santa Margarita, 1941. Photo courtesy of Jorge M. Rolland C.

FIG. 40. Tolentino, Ana María, Rolland, Martha, and Catherine Rolland, the wife of Rolland's son Jorge, Rancho Santa Margarita, 1941. Photo courtesy of Deanna Catherine Wicks.

FIG. 41. Rolland's grandchildren playing in a stream bordering Rancho Santa Margarita, 1947. Photo courtesy of Deanna Catherine Wicks.

FIG. 42. Pres. Manuel Ávila Camacho and Rolland discussing the City of Sports, 1944. Photo courtesy of Jorge M. Rolland C.

FIG. 43. Officials visiting the construction site for the City of Sports, 1944. Photo courtesy of Jorge M. Rolland C.

FIG. 44. Construction of the Plaza de Toros, ca. 1944. Photo courtesy of Jorge M. Rolland C.

FIG. 45. Aerial view of the City of Sports under construction, ca. 1945. Photo courtesy of Jorge M. Rolland C.

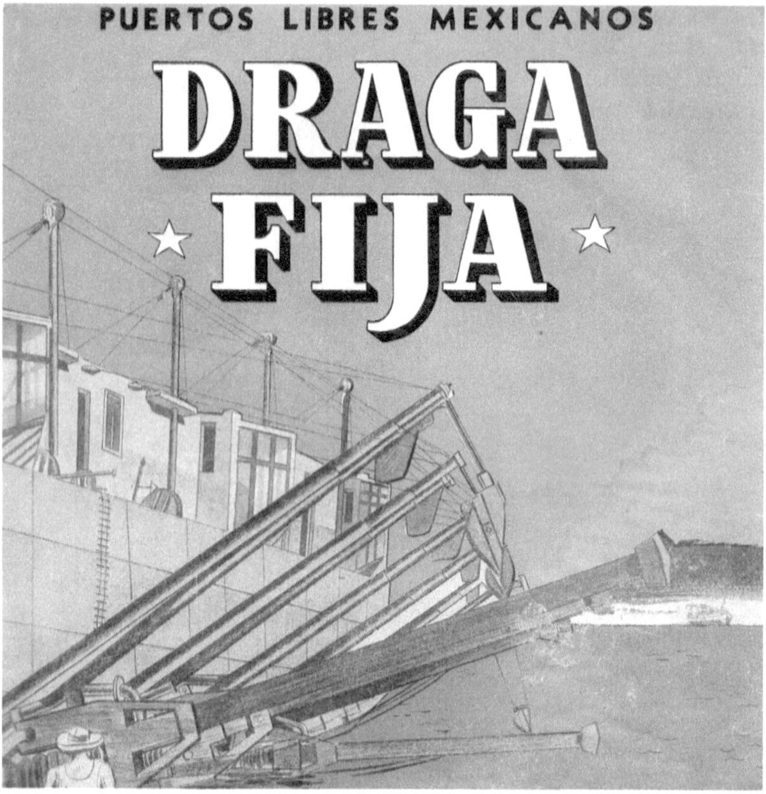

PUERTOS LIBRES MEXICANOS
DRAGA FIJA

FIG. 46. (*opposite top*) The fixed dredge, as shown on the cover of the free ports promotional booklet *Draga fija*, 1950. Used with permission by Jorge M. Rolland C.

FIG. 47. (*opposite bottom*) Rolland with workers and a foreign specialist at Salina Cruz, ca. 1950. Photo courtesy of Jorge M. Rolland C.

FIG. 48. (*above*) Rolland family photo, Mexico City, 1952. Photo courtesy of Jorge M. Rolland C.

FIG. 49. Rolland family photo, Córdoba, 1959. Photo courtesy of Jorge M. Rolland C.

FIG. 50. Rolland, Tolentino, and family at a church event celebrating Rolland's eightieth birthday, 1961. Photo courtesy of Deanne Catherine Wicks.

10

The Undersecretary

WHEN THE DECEMBER 1, 1938, issue of the magazine *Clave* hit the stands in Mexico City, it included a letter that consisted of a scathing diatribe against Modesto Rolland, the "modest engineer" and "inventor of a famous tortilla machine and a way to make Mexican soil more productive by hanging basketfuls of earth in the trees for planting potatoes." *Clave* was a new Marxist publication edited by artists and intellectuals, including Diego Rivera, and associated with Mexico's recently arrived preeminent exile, the Soviet revolutionary Leon Trotsky. The letter's authors included Rivera, along with his fellow artist and companion Frida Kahlo, the young and on-the-verge-of-being-famous writer Octavio Paz, the iconoclastic but far-from-communist Salvador Novo, and dozens of other figures prominent in art and intellectual circles.[1]

This sore kaleidoscope of artists was protesting Rolland's order to have the artist Juan O'Gorman change his mural at the Mexico City airport. It promoted Marxism and portrayed fascist leaders as bestial demons. While Rolland did indeed continue to work on various contraptions, "the modest engineer" had also become undersecretary of communications and public works. But according to his detractors, he was not directed by Pres. Lázaro Cárdenas (1934–40) or even his immediate superior, Gen. Francisco Múgica. Rolland's hand was moved, in their view, by Hitler and Mussolini.[2]

Kahlo and company had a hard time understanding how Cárdenas and Múgica, seen by many Mexicans as defenders of the Left, could allow a probusiness and art-censoring engineer like Rolland to hold such a prominent position in the government. The letter writers failed to grasp Cárdenas's more pragmatic side and his broader ideas about government and development. And although these artists marched, painted, operated presses, and debated passionately during this turbulent time, they seem not to have understood clearly the complexities that drove the decision to censor O'Gorman's mural. That, or their letter, was hyperbole.

Rolland was no fan of communism, but neither was he a henchman of Hitler. The reality was that Rolland was acting on behalf of his political superiors in a way that he perceived to be in Mexico's best political and economic interests. Earlier that year Cárdenas had expropriated foreign oil companies for abusing workers and disobeying a Supreme Court ruling. The decision was backed by most Mexicans, including Kahlo and Rivera. Many American and British business operators and government officials protested, calling for a boycott of Mexican petroleum. And although Cárdenas backed left-leaning Republican Spain against the fascists during the ongoing Spanish Civil War (1936–39), he was not above selling oil to Italy and Japan to maneuver around boycotts and diversify Mexican economic partners. Cárdenas and Rolland differed widely on politics, but they shared a sense of pragmatism and a desire for industrial progress.

The Cárdenas years were exciting times, both at home and abroad. It was clear during Cárdenas's campaign for the presidency that he would tilt the government back to the left. He lashed out at "bourgeois pseudo-cooperativism," calling for increased land redistribution, worker organization, and collaboration between the people and the state.[3] He would go on to defy expectations by ending Calles's unyielding influence over politics, forcing the former president into exile in 1935. Cárdenas granted *ejidos* at a pace that far surpassed that of his predecessors, he backed worker strikes for better conditions, and he reformed the Partido Nacional Revolucionario (PNR), transforming it into the Partido de la Revolución Mexicana (PRM), creating an even more corporatist structure that included peasants, urban laborers, the military, and even white-collar bureaucrats. All of these changes met with resistance, but except among most industrialists and certain local *caudillos*, he was an extremely popular president, talking to peasants and laborers with some regularity. Cárdenas genuinely desired to improve living standards of everyday Mexicans, but his call for increased worker participation was also driven by another motive—to expand Mexican industry, consumerism, and modernity by creating a more interventionist state.[4]

Mexico found itself in a fast-changing world of intensifying international conflicts. In many places the Great Depression dragged on. Getúlio Vargas had risen to power as a dictator in Brazil, putting in place his own corporatist government. Tensions were rising over global capitalism and global communism, foreshadowing the Cold War. Italian soldiers invaded Ethiopia in 1935. The Spanish Civil War pitted a motley conglomeration of leftist organizations against the fascist general Francisco Franco. Mexico was one of the few official supporters of the Republican or anti-Franco forces; it sent ammunition and took in refugees. Intertwined with this war was the rise of fascism in other parts of Europe, especially in Italy and Germany. Germany annexed Austria amid only weak protest in 1938. The year before, the rightist government in Japan had invaded China.

Many foreigners saw Cárdenas's rise to the Mexican presidency as a symbol of renewed protest against foreign enterprises. And there were serious clashes with a number of foreign transnational corporations, most notably in the oil industry. But Cárdenas stated early in his presidency that he aspired to work in "mutual cooperation" with other countries as much as possible.[5] Intelligent observers recognized that although Cárdenas had moved the Mexican government leftward, he was not as radical as some of its fiercest critics often claimed. He was not attempting to build a center of global communism but a "middle-class capitalism . . . in the name of something like socialism."[6]

Rolland obtained high-ranking positions in Cárdenas's government, serving at different times as the director general of railway construction, undersecretary of communications and public works, undersecretary of the national economy, and general manager of the free ports.[7] Rolland's presence was one of many signs that the new president was distancing himself from Calles. At first Rolland carried on in his work of managing the construction of railroads, especially in the southeast, though he moved on to influence other developments in ports, communications, transportation, petroleum, and agriculture.

Rolland's infrastructure-focused work under Cárdenas highlights further the importance of infrastructural projects to the goal of nation building. In order to expand education, create a greater sense of *mexicanidad*, or Mexicanness, and exploit the nation's natural resources, Mexico needed more roads, railroads, storage facilities, radios, telephones, airfields, and functioning sea ports. Cárdenas especially hoped to incorporate more peasants and industrial workers into the state apparatus as a means of driving industrial development in the face of the Great Depression, which had temporarily decreased access to manufactured goods from the United States and Europe. These infrastructural projects, so the thinking went, could bind together different cultures, spread education, provide wage-paying jobs and appease unions, improve government relations with industrialists, and expand the organizational capacity of the state. For Rolland and other engineers and architects these material pathways were crucial to class conciliation, social unification, and progress. Technocrats like Rolland were the Mexican nationalists most capable of designing these binding ties.

New Administration, New Opportunities, Old Friendships

A critic of ejidos, Rolland may at first seem to be a strange selection for undersecretary of communications and public works. He disagreed with Cárdenas and Múgica on issues of labor and agrarian reform. Rolland detested Cárdenas's populist approach, but it was not uncommon to have high-ranking officials in Cárdenas's ministries differ significantly in their politics and worldviews. As one historian put it, "the Cárdenas presidency was marked by a tense, at times even conflictive, coexistence among radically anti-imperialist, moderately nationalist, and even conservative, openly pro-business policy makers."[8] Rolland borrowed a bit from each of these categories, but his differences with Cárdenas were not so numerous that the president couldn't find uses for him.

Rolland also owed his position to friendships and changing alliances. Múgica, serving as Cárdenas's secretary of communications and public works, had a long and friendly history with Rolland. Múgica had been a long-serving revolutionary general, former governor of Tabasco, and longtime ally of Cárdenas's. As far as top politicos went, Múgica was radical. Múgica promoted agrarian reform and secularism as governor. And although Cárdenas himself was not planning a communist utopia, Múgica admired the writings of Vladimir I. Lenin, believing that the Mexican government needed to carry out a program of state-driven capitalism before Mexico could progress to a true socialist society. Múgica ardently opposed those people and organizations he saw as exploitative imperialists. As head of communications and public works, he strove to construct new roads and to make electronic communications less of a luxury—to extend their use to everyday people, putting in place "public service socialism."[9] He battled transnational telephone providers in an attempt to put the industry under greater state control, something that Rolland had promoted in decades past.

Múgica and Rolland's relationship dated back to Veracruz in late 1914, thus exhibiting once again the importance of the Constitutionalist conference that took place there in December of that year. In 1921 they had kept in communication about the difficulties of land redistribution.[10] In the mid-1920s Múgica had worked as a paymaster for Heriberto Jara, overseeing most of the finances for Xalapa Stadium. Múgica appears to have respected Rolland's talents, nationalism, and drive to modernize Mexico. And although Rolland and Múgica differed on political matters, Rolland was not inherently opposed to state-directed capitalism and social programs; he had after all played a substantial role in realizing them.

Múgica and Rolland were also both mutual friends of Jara. Cárdenas had brought Jara, another political adversary of Calles, into the government as inspector general of the army in 1935. Four years later Jara became president of the newly formed PRM. Rolland's past friendships with Jara and Múgica, as well as his adversarial relationship with Calles, made him, despite political differences, a shoo-in for the Cárdenas bureaucracy.

Ministry of Communications and Public Works

Within the federal government Rolland continued to focus on his aspirations for southeastern Mexico and Tehuantepec in two high-ranking positions within the Ministry of Communications and Public Works: director general of railway construction and undersecretary. In the former role he oversaw developments in the port of Progreso and supervised advancements made at the shipyard in Campeche.[11] He also oversaw the construction of parts of the Southeastern Railroad. The ministry made substantial progress on the railway. A band of engineers and an army of workers, including military brigades, cleared paths through dense jungles with axes, tractors, and fire. They began constructing bridges over the rivers and swamps of Campeche, Tabasco, Chiapas, and southern Veracruz. Rolland was a man possessed by his vision. He was certain that the railway would be completed by 1940, which would mean realizing his goal of seeing the Yucatán peninsula connected to Tehuantepec and Mexico City by rail. In his zeal he underestimated nature and overestimated the rate of progress. Facing immense geographic difficulties, "copious rains," and increasingly limited resources in the face of World War II, the railroad would not be completed until 1949, well after Rolland had left his position. There was, however, limited train service on the parts of the line that had been completed by 1940.[12]

Rolland also branched out into other areas of development. As undersecretary of communications and public works, a position he obtained in early October 1938, Rolland did a lot of rubber stamping.[13] He approved radio fines and licensing fees, something he had earlier protested in the 1920s. He inaugurated the construction of new buildings at the Guadalajara airport.[14] Rolland's name is on a number of documents establishing committees to study conservation, to help organize and keep track of cooperatives, and to provide better conditions for chicle and hardwood harvesters in southern Mexico.[15] Put alongside his 1930 discussion of poor forest management in Morelos and his call to better protect resources in Baja California in 1939, Rolland appears to have had genuine concern about the environment and wanted to protect it, not as a way to halt development but to better sustain Mexico's resources and economy.[16] He signed on to decrees that oversaw the expropriation of the oil industry, which involved establishing in various states councils that consisted of government officials and union leaders tasked with taking over petroleum operations.[17]

As discussed earlier, Rolland also condemned public artwork that he deemed unacceptable to the goals of the Cárdenas administration. The O'Gorman mural episode arose from the aforementioned domestic and international tensions. It represented a limit to the government's tolerance of artistic expression based more on economics and realpolitick than a desire to quash O'Gorman. And Rolland, "Mr. Bothersome," more than Tata (papa) Cárdenas and radical Múgica, was the man best suited to carry out the controversial decision. He took the fallout, leaving his superiors' leftist bona fides less tarnished. But it was becoming clearer that the Cárdenas administration was gravitating back to the political middle, a moderating reaction to the backlash over the oil expropriation, pressure from the United States, and the growing influence of more moderate members of the PRM.

Prominent people from the far corners of Mexico's art world lambasted Rolland. After stating that Rolland might not be able to handle his undersecretary post and thus giving no credence to his prior experience in government positions or in the revolution, Rivera and the other signers of the *Clave* letter called Rolland a capitalist enemy of the worker who in the name of the state "tramples on the most elementary rights of self-expression[,] ... who behaved with vandalism, typical of totalitarian police power, when he destroyed the Central Airport paintings of a Mexican artist who supported working-class interests."[18] And although there was much in the angry letter of Kahlo and company that was overblown and inaccurate, its authors were right that Rolland emphasized order and supported capitalism. The days of his bombastic tirades against the evils of foreign financiers faded further into his past.

To better understand Rolland's actions, though, they have to be placed in the context of his personal pursuits and the foreign policy concerns of the Cárdenas administration. Rolland was particularly concerned with O'Gorman's inclusion of a famous quote from Karl Marx—"With the Communist revolution the proletariat has nothing to lose but their chains, and in exchange they have the world to gain"—as well as the artist's negative portrayals of Mussolini and Hitler. Rolland had developed a distaste for worker radicalism and the influence of communism. But Rolland was driven even more by practical considerations. Even before the oil expropriation, the Cárdenas administration had worked on expanding Mexico's oil market to more countries besides the United States, a plan that received increased attention after the nationalization of the industry. As the historian Stephen Niblo has noted, U.S. corporations pushed the Mexican government toward the Axis powers when the former painted Mexico as an "outlaw nation" for nationalizing U.S. oil facilities.[19] At the time when O'Gorman was painting his mural in the Mexico City airport, the Cárdenas administration was working with German, Italian, and Japanese representatives to sell oil to their governments.[20] These officials regularly came through the airport.

Japanese Oil Interests

Rolland was directly involved in Japanese oil endeavors in Mexico. He had long desired to open more Mexican trade with Asia, something he thought would improve his home region of Baja

California and benefit Mexico as a whole. In 1934 and 1935 two Japanese entities set up major petroleum operations under Mexican shell companies: Compañía Mexicana de Petróleos "La Laguna" and Compañía Petrolera Veracruzana. Rolland was a shareholder in La Laguna, and he also briefly served as president and general manager of the latter company. The companies drilled mostly in Veracruz but also in Tamaulipas and Tabasco. When Cárdenas expropriated the foreign oil companies in 1938, these operations went untouched. The man most associated with them, Kisso Tsuru, was a naturalized Mexican citizen, and he had gained significant influence in Cardenista circles. By 1938 Rolland was no longer managing the company, but he and a handful of bankers and military leaders remained important shareholders.[21] In reality the businesses were mostly Japanese, but in the eyes of certain government leaders the Japanese gave more respect to Mexican officials and represented a potentially important economic outlet. This relationship was also a clear example of crony capitalism.

Japanese officials toyed with buying large quantities of oil from Mexico throughout the 1930s. Rolland and Múgica pushed them to do so. There were debates between Japanese business operators, the Japanese navy, and Japanese political leaders about whether stocking up on Mexican oil was a good idea. Oil was crucial to their attempt to further industrialize and to expand their influence and territorial possessions in Asia and the Pacific. But in the mid-1930s Japanese officials ultimately turned down plans to invest further in Mexican petroleum, arguing that is was more expensive and of poorer quality than petroleum from the United States, which continued to sell oil to Japan until the end of the decade. By late 1939, however, the Japanese government reversed its decision as it drove its military farther into Asia and foresaw difficulties getting oil from the United States. Japanese political and business leaders backed drilling and further oil exploration in Mexico and suggested that they would make massive purchases of Mexican crude, kerosene, and gasoline in the near future in exchange for the Mexican government upping its importation of Japanese rayon, a cellulose-based textile fiber. Rolland, while serving as undersecretary of the national economy, signed off on a contract allowing the Compañía Petrolera Veracruzana, the company of which he had earlier been president, five-year exploration rights to a large swath of Veracruz along the Tehuantepec Railway. Another Japanese stipulation for any major oil deal between Japan and Mexico was that a pipeline be built across Tehuantepec and that the Mexicans modernize the port of Salina Cruz, which had fallen on hard times since Álvaro Obregón shut down the free ports in late 1924.[22]

Rolland had a number of reasons to work closely with Japanese investors. Reflecting on Rolland's interaction with these Japanese firms, the historian María Emilia Paz Salinas has chalked his actions up to simple corruption. She writes that the Japanese worked with "corrupt Mexican officials," and she specifically calls Rolland "the corrupt undersecretary of communications."[23] Surely Rolland didn't throw away the money that came from his shares and his position with the Japanese oil companies. In his frustration with the slow pace of Mexico's progress he participated in crony capitalism. Considering his loud longtime criticism of corruption and personalist politics, his actions here appear hypocritical. Some observers contended that he was taking bribes to give the Japanese preferential treatment. William B. Richardson, a U.S. banker in Mexico City, stated that Rolland and Gen. Juan Barragán received 800 pesos monthly "for obvious reasons."[24] Lines between political and economic elites became blurred during revolution-era reconstruction, and graft was commonplace. Government officials before, during, and after the revolution also frequently sat on the boards of foreign corporations to serve as intermediaries between the government and foreign financiers in an attempt to balance Mexican sovereignty with the need for foreign capital and technology.[25] Whether Rolland was taking direct bribes or not, he definitely used his position to his own profit, even if he thought he was benefiting Mexico in the process.

But Rolland most definitely was an intermediary between

these Japanese entities and the Cárdenas administration, which was navigating probable U.S. and British embargoes over the nationalization of those countries' petroleum industries in Mexico. Múgica and Cárdenas knew what Rolland was doing. Múgica in particular backed the joint Japanese-Mexican adventure, but there were divisions between ministries about this endeavor. Secretary of Finance Eduardo Suárez opposed Rolland's actions, especially as tensions between the United States and Japan increased, and he argued that the project was too dangerous; he feared provoking the U.S. government. Cárdenas appears to have sided with Múgica and Rolland at first, though he quickly pulled his support due to a shifting political situation during his last year in office.[26]

The United States began to place significant pressure on Mexico after it came to light that Rolland had signed off on the potentially lucrative contract for the Compañía Petrolera Veracruzana. Alarmed and "disgusted" U.S. officials claimed incorrectly that Rolland still "presided" over the Japanese operations while approving the concession as undersecretary of the national economy, a falsehood further spread by the *Daily Record*, an English-language newspaper in Mexico City.[27] Ambassador Josephus Daniels threatened "serious harm" to Mexico if the agreement was not annulled.[28] U.S. agents were unclear of the exact details of the Japanese-Mexican oil trade, but they knew that there had been increased collaboration. Officials within the Mexican government were divided about the concession, but with tensions between the U.S. and Mexican governments otherwise easing, Cárdenas had the contract repealed.[29] The incident spurred more drastic changes as well. The Cárdenas administration, in solidarity with the United States and the "embargo on war materials to the totalitarian countries," announced that it would stop all shipments of petroleum, quicksilver, manganese, and scrap iron to Japan.[30]

For Rolland, the Japanese had provided a potential path for him to realize his own personal long-term aspirations for Mexico. That is what drove him. The Japanese companies allowed for petroleum development without Mexico relying solely on the United States and Britain, countries with petroleum enterprises protesting Mexican policies and trade. Rolland had been trying to develop Tehuantepec and to move Mexican commerce into the Pacific for more than two decades. Rolland and other Mexican, Japanese, and U.S. government officials were well aware that petroleum sales to Japan meant revamping Tehuantepec and the connecting ports.[31] In 1939 Rolland, together with Francisco J. Aguilar, the ambassador to Japan, stressed the need to increase trade to Japan at the Congreso Nacional de Exportadores, or National Congress of Exporters. This trade, Rolland pleaded, depended on Mexico's ability to bring the free ports back to life, to simplify taxes, to calm worker unrest, and to build a better merchant marine.[32]

Beginning Anew with Old Plans

Cárdenas meanwhile reopened the free ports and put them back under Rolland's management. The president believed in Rolland's vision for expanding trade, increasing infrastructure, and developing Mexico's southern and westernmost territories. Bureaucratic shakeups in the lead-up to the 1940 election also influenced Cárdenas's decision. In January 1939 Múgica resigned his position as secretary of communications and public works in order to run for the presidency. Many people thought Múgica, a leftist and close ally of Cárdenas, was the most likely candidate to obtain the support of Cárdenas and the PRM. Melquiades Angulo, who like Rolland was an engineer and who had served as undersecretary before Rolland, took Múgica's place.[33] Rolland and Angulo had both served as director of National Railways. Like Rolland, Angulo was more conservative than Múgica. Rolland and Angulo, however, did not get along.

It is unclear where their distaste for each other began. It probably had something to do with the fact that Rolland had replaced Angulo as undersecretary the previous year. Angulo had reportedly left the position because of differences with Múgica.[34] Rolland told Angulo that if they were to work together they would have to let bygones be bygones, to "erase the bad impressions as soon as possible" in order to further the ministry's work and to aid the president. Rolland stressed that he was "energetic, apolitical, and honest" and just wanted to "quietly work" on his many obligations.[35]

Despite Rolland's talk of making amends, the situation never improved. Rolland attempted to set up an appointment to meet with Angulo, who then failed to show up. Angulo never spoke with Rolland for his entire first month as secretary. This, to Rolland, was inexcusable. Rolland was now serving as both manager of the free ports and undersecretary of communications and public works. He had "great responsibilities." The most important issue was that the ministry was a mess because people did not know to whom they should turn. All of this, Rolland concluded, was because Angulo still held a grudge.[36]

Angulo did eventually get to work. Many of his actions focused on reversing his predecessor Múgica's policies toward international corporations. For example, Angulo attempted to make amends with foreign telecommunications companies, something that Rolland, despite his differences with Angulo, publicly went along with.[37] In general, Angulo promised closer relations with foreign and domestic businesses and provided reassurances that the Cárdenas administration would not nationalize more industries.

Cárdenas was moderating his policies, hoping to cool the backlash over the oil nationalization, to improve relations with the United States, and to increase foreign trade and investments. Instead of backing Múgica for the presidency, Cárdenas threw his support behind the more moderate Manuel Ávila Camacho, who had been serving as secretary of national defense. Cárdenas retained Angulo as head of communications and public works and made Rolland both manager of the newly reopened free ports and undersecretary of national economy. Cárdenas supported Rolland's determination to "bring in outside capital without compromising our [Mexican] prerogatives and laws."[38] In these positions Rolland faced less resistance, worked more freely on developing Tehuantepec and western Mexico, and pivoted back to working with U.S. business leaders.

Rolland's continuing desire to revamp transportation across Tehuantepec stemmed from his ongoing obsession with increasing development in western Mexico, especially in the Baja California peninsula. And he was not alone in arguing that the frontier regions of Mexico needed better development if Mexico was to thrive as a whole. Rouaix had also called for Baja California to receive more attention.[39] But Rolland worked harder and in more diverse ways than Rouaix to bring the peninsula to the attention of the Cárdenas administration. Rolland pushed for expanding free trade. He promoted the use of guano from islands in the Pacific and in the Sea of Cortez to fertilize crops. He brought to light opportunities for chemical production, especially the existence of large, unexploited deposits of sodium carbonate—used in the production of glass, caustic soda, and as a base in a number of chemical processes—in a number of lagoons of Baja California. Múgica had pleaded with the president to pay attention to Rolland and to make sure that other departments did not sign off hastily on any concessions for mining the resource because the state could use it for its own attempts to build a stronger chemical industry. Múgica would continue to work with Rolland when the former became governor of the Territory of Baja California Sur under Cárdenas's successor, Ávila Camacho (1940–46).[40]

During his transition to the Free Ports Commission and the Ministry of the National Economy in the spring of 1939, Rolland joined Cárdenas and a number of other high-ranking government officials on a month-long tour of western Mexico, including the Baja California peninsula. The president summoned Rolland to meet him in Hermosillo, Sonora, to discuss the Kansas City Railroad, free ports, and the free-trade zone in the northern parts of Sonora and Baja California.[41] The entourage, which also included a number of reporters, boarded a boat, the *Guanajuato*, to tour the Pacific coast.[42] Rolland's goal was to further influence Cárdenas to invest more resources in the region, to increase the number of free ports, and to expand the free-trade zone that the president had already established in the Territory of Baja California Norte to include the Territory of Baja California Sur.[43]

In early 1940 Rolland, creeping closer to his sixtieth birthday, published an account of his tour with Cárdenas in the magazine *Novedades*. While John Steinbeck was waxing poetic about life, mysticism, and the past with the biologist Ed Ricketts on their Gulf of California expedition, Rolland pleaded with readers of *Novedades* that they needed to reenvision Sonora and Baja California. The region was not, he declared, an unproductive desert wasteland or at least it did not have to be. To the contrary, Rolland argued that it could be one of the most productive areas in Mexico with the proper attention. He wanted to make Puerto Peñasco in northwestern Sonora a free port in order to expand trade via the Kansas Railroad and, via a new road, into Baja California. He foresaw the blooming of an agricultural renaissance in the Baja California territories, as had happened in the U.S. state of California following "reclamation" projects there. There were many places on the peninsula, Rolland continued, that had excellent soil, a wonderful climate, and a decent amount of underground water. There were places perfect for the production of wheat, cotton, grapes, and dates. What the peninsula still needed was cheap and efficient irrigation and more roads.[44]

All of this tied back to Tehuantepec and Rolland's work during the revolution. The best way to develop irrigation in Baja California, according to Rolland, was to ship petroleum from the Gulf regions across Tehuantepec to Salina Cruz. Apparently he had given up on his wind-powered, more ecologically sensitive pump. Instead he pushed for the construction of a refinery in Salina Cruz to produce gasoline to fuel generators that could cheaply power water pumps. By this means Mexicans could "conquer the desert."[45]

He also called on the Mexican government to make the southern half of the peninsula into a free zone, like the Territory of

Baja California Norte.⁴⁶ Rolland, alongside his colleague Ulises Irigoyen, backed the idea of turning much of the border into a free zone and extending it across the entire Baja California peninsula. Rolland had long been a proponent of free trade, but he appears to have been influenced even further by what he saw as a bloated and ineffectual government. He wrote that extending the free zone would create "innumerable opportunities" that would arise from a less burdensome tax structure and from cutting back an "immoral, sterile, and useless bureaucracy."⁴⁷

While he promoted cutting bureaucratic red tape, Rolland, along with other prominent engineers, agronomists, and Bajacalifornianos, established the Pro-California Commission. The group lobbied Cárdenas to put more energy into developing the peninsula responsibly, into planting trees and fruit-bearing orchards, and into sustainably using and conserving marine resources, all while liberalizing trade.⁴⁸ The renowned engineer and conservationist Miguel Ángel de Quevedo of the Pro-Tree Committee worked with Rolland and the Pro-California Commission on issues regarding forestry.⁴⁹ To Rolland, developing and protecting Baja California were both patriotic duties: "All Mexicans are obligated to think of the problem of Baja California and aid with affection and love in its development, since a chain is no stronger than its weakest link."⁵⁰ As shown by Cárdenas's tour with Rolland, the president gave Baja California and Rolland's ideas serious attention. The two men remained in direct contact, creating plans to place the peninsula at the forefront of the next presidential administration's agenda and working to obtain coconut seeds from Isla Cozumel and Colima to plant at El Mogote, a small peninsula jutting out into the Sea of Cortez opposite La Paz. Rolland also pleaded with Cárdenas on behalf of the Pro-California Commission to halt certain types of net fishing from boats in the Gulf of California, which he rightfully argued was immensely damaging to numerous marine species.⁵¹ Rolland telegraphed Cárdenas about building roads and a tourist industry and about how to best take advantage of free-trade zones and their associated reformed tax codes.⁵²

At first glance Rolland's arguments may appear contradictory. He called for less government interference while at the same time working as a government agent who pushed the federal government to take direct action in developing and protecting resources in regions he believed needed greater attention. Rolland promoted campaigns to increase municipal independence while attempting to connect Mexico's towns and resources to the central government. In his mind it was possible to strengthen local-level democracy, private business initiatives, and the central state all at the same time—to him the processes were reliant on each other. He wanted less government red tape for Mexican entrepreneurs, but he still felt it necessary for the Mexican government to build a stronger infrastructure, to regulate certain practices, and to negotiate and supervise foreign capitalists. To him these practices were complementary, not contradictory. The road to success was a wise mixture of socialist and capitalist practices.

Rolland saw the free ports as the best means to bring greater development to western Mexico via trade liberalization, but their reinstatement faced a number of difficulties. By the end of the 1930s the ports were in a deplorable state. Of the eighteen cranes that had operated in 1924, only two were still functional. Most of the warehouses needed major repairs. The docks of Coatzacoalcos, formerly Puerto México, were in shambles. In Rolland's words, "Ruin is paramount everywhere and even the town itself is raggy." Salina Cruz had also fallen into disrepair. The power plant was no longer functioning, and the entry channel to the inner port had filled with sand. It was now impassable for most ships. The one positive was that the Tehuantepec Railway still functioned, even though it needed some work on sections of its track. In the last year of the Cárdenas administration workers began repairing warehouses and dredging the entry channel at Salina Cruz.[53]

Rolland also needed the support of U.S. backers. Cárdenas had been working to repair U.S.-Mexican relations, and U.S. representatives and military officials began to take a renewed interest in Tehuantepec as American participation in World War II increased. Some members of Congress in Washington DC had begun to call for ending tariffs and opening free trade between American countries as a way to economically challenge Germany, which looked like it would dominate much of the European continent.[54] In light of these changes and Rolland's recent collaboration with Japanese oil producers, Rolland and Irigoyen, his close and talented collaborator, went to great lengths to stress that the new free ports initiative would be carried out in association with the United States. Rolland even appointed free port representatives to be stationed in San Francisco, New Orleans, and New York.[55]

In a speech to the Sociedad Mexicana de Geografía y Estadística, or the Mexican Geographic and Statistics Society, Rolland capitulated, stating that Mexico could not fight the United States or the most powerful capitalist interests of the world: "We cannot fight the U.S.A. on account of a problem that they consider of the greatest importance to their welfare [the commercial and strategic value of the Isthmus of Tehuantepec].... We must face the cold facts and not waste energy on useless lamentations or trying to enter an uneven fight." Mexico would have to come to terms with the United States, but the United States would need to build a cooperative agreement in which Mexico's "nationality and rights remain unimpeded."[56] Although Rolland remained an ardent fighter for Mexican sovereignty, he had given up fighting against collaboration with U.S. officials and corporate leaders. Rolland feared that if Mexicans did not work with the United States and the corporations based there, then the U.S. government might take the isthmus by force, turning it into another Panama Canal. Rolland's advice was to make the United States a deal that made direct intervention unnecessary, giving Americans preferential treatment but allowing Mexico to keep its territorial sovereignty intact. If done wisely, Rolland suggested, the partnership could ultimately be beneficial to both the United States and Mexico.

In essence Rolland called for policies that were similar to the practices put in place under the former dictator Porfirio Díaz, which had provided American business leaders with greater access to Mexican resources as a means to help develop Mexican infrastructure but also as a way to silence American calls for further U.S. annexation of Mexican territory.[57] The Mexican government would have to carry out a balancing act that weighed Mexican development and sovereignty against the desires of its powerful neighbor. Mexico had changed significantly since 1910, but some policies had come full circle.

11

Going Big

ON DECEMBER 18, 1940, not even a month into the presidency of newly elected Manuel Ávila Camacho, the attorney general "opened an investigation of alleged acts of maladministration committed in favor of a Japanese company by Modesto C. Rolland."[1] Rolland was accused of providing Japanese interests a permit to import seven thousand tons of rayon while at the same time denying permits to other rayon importers. He was charged with signing this concession as undersecretary of the national economy. This monopoly allegedly allowed the Japanese to mark up their prices by 300 pesos more per ton than the usual price. According to one newspaper, the racket was broken up when the administration changed.[2] The new president promised to clean up the government in a "moralization campaign."[3]

Rolland's relationship with the Japanese became more exposed at a terribly inopportune time. Alliances in the escalating world war were hardening, and Japan had become an enemy in the eyes of most of the U.S. public. The Mexican government, increasingly aligned with the United States and formally so after the Japanese attack on Pearl Harbor, did not want to look to be supporting the Axis powers. Times had changed in the few short years that Rolland had worked with Japanese business operators in Veracruz.

The entire "moralization campaign" was more show than anything else. Ávila Camacho promised to weed out corruption and put in place "an honorable government with honest employees and functionaries."[4] But Ávila Camacho was a friend of Lázaro Cárdenas. His family, power brokers in the state of Puebla, was not known for its purity. His brother was the poster child for violence and graft among revolutionary officials. And, unbeknown to people at that time, the 1940s would go on to be seen as an era of widespread government corruption.

Rolland claimed he had done nothing wrong. He sent letters to the press and powerful officials. He argued that the price for the rayon had been fixed by the Ministry of the National Economy and that the president had himself authorized the importation of five million to ten million kilograms of the product. The Ministry of Finance was also aware of Cárdenas's orders and had been involved. Rolland contended that this "dark matter," his phrase for the allegations, had been brought about by a consortium of foreigners involved in the rayon trade who were influencing certain government representatives.[5]

Rolland survived the accusations. After the initial outcry among members of the U.S. and Mexican press, there were no other public reports on the matter. Apparently the president found Rolland honest enough to retain him in his position after all. The dearth of evidence makes it difficult to ascertain the reasons why Rolland was let off the hook or if the accusation had any merit. Perhaps he was a target of political sabotage. Rolland lambasted other officials for corruption and defended himself against the accusations. What evidence about Mexico's rayon trade with Japan that does exist indicates that Cárdenas knew about the deal and Rolland's role. Other evidence, however, points to possible abuses of power, if not in regard to the rayon deal then in the land deal Rolland had signed off on for the Compañía Petrolera Veracruzana. At the very least the land deal was an unwise decision that was bound to fuel allegations and backlash. It didn't look good.

Out of the shadow of corruption accusations, Rolland set out to undertake his most ambitious works—continuing to expand on his ideas for the free ports of Tehuantepec and taking a leadership role in a commission to make infrastructural changes to the southwestern state of Chiapas. Rolland designed a massive multitrack railway to carry oceangoing ships across the Tehuantepec isthmus. He also started work on a stationary dredge that was to keep the inner port of Salina Cruz from filling with sand. Rolland also bought a ranch in the mountains of Veracruz and planted a large orchard there. In Mexico City he built the largest bullfighting stadium in the world, part of a large development called the City of Sports.

The increasing scale of Rolland's undertakings mirrors that of the governments of the 1940s. Often incorporating World War II rhetoric espoused by the Allies, such as equating production not only with the revolution but also with the defense of democracy in the face of rising fascism, administrations of the era focused heavily on state-directed industrialization, economic expansion, infrastructural development, and big projects.[6] The Ávila Camacho and Miguel Alemán Valdés (1946–52) administrations provided tax incentives for certain equipment and foreign companies—U.S. direct investment increased significantly—but in general these administrations pushed protectionist plans for development, raising tariffs and applying import controls in an attempt to further spur what economists have termed import substitution industrialization, or ISI.[7] Ávila Camacho and Alemán carried out these policies in an attempt to boost domestic industry and consumption, especially as World War II decreased access to manufactured goods from foreign markets. These governments also supported Rolland's free ports, showing that they were open to different approaches to spurring the economy.

Despite the support, Rolland constantly ran into problems. He overstretched himself, warred with competing agencies and an old friend-turned-enemy, struggled with workers, and faced pressures caused by the outbreak of World War II, including increased scrutiny, greater U.S. demand for Mexican resources, and limited foreign interest in using the ports as manufacturing centers. Rolland's ambitious plan to build a multitrack "ship railway" across Tehuantepec initially found strong support, but he struggled to find the required private-sector financing from the United States. It also proved more difficult to build his massive automatic dredge in Salina Cruz than he originally anticipated.

The most successful project Rolland undertook turned out to be the two stadiums he included in the City of Sports. Exemplifying the increased interest in spectacle during the 1940s, the stadiums were grand and privately financed by a wealthy entrepreneur. Unlike the free ports, there were no competing bureaucracies, less red tape, and no problematic international affairs. In charge of the design, well funded, and largely free to enact his vision, Rolland carried out the project with efficiency and skill.

The Free Ports Become a Reality

Whatever Rolland's culpability in the rayon scandal, Ávila Camacho and Rolland came to terms. Rolland's dealings with the Japanese may have strained U.S.-Mexican relations, but Rolland's willingness to play ball with U.S. interests seems to have aided his political survival. The president realized that Rolland was not only an intermediary between Japanese interests and the Mexican government but also a competent broker between the Mexican government and U.S. businesses interested in the strategically important Isthmus of Tehuantepec.

Rolland respected the Japanese men he had partnered with, but he always aimed to work with whomever he thought he needed to in order to advance his vision for Mexico. Even while collaborating with the Japanese, Rolland reached out to officials in Washington DC, recognizing that the free ports would cater more to American shippers than anyone else. He also knew that Tehuantepec had taken on an even greater strategic importance because of World War II. The U.S. government reopened consulates in Coatzacoalcos and Salina Cruz.[8] Rolland was not blind to the limitations placed on Mexico by the ever-present realities of outside forces. His pivot back to the United States was quick and nearly total.

Rolland, with the scandal behind him, continued working on the free ports. The same newspapers that had only months before called him a corrupt bureaucrat now praised him as the "Apostle of the Free Ports."[9] The government slowly funneled millions of pesos into repairing the Tehuantepec Railway, as well as dredging the channels and repairing the docks and warehouses at the port of Salina Cruz. It finally appeared that the on-and-off again support had turned into more reliable backing and that real progress would be made.[10]

U.S. officials concerned about the likelihood of war with Germany and Japan supported the renewed efforts to improve transportation across the isthmus. Eager to find ways to relieve congestion in the Panama Canal, the U.S. embassy and State Department officials kept close tabs on the progress made in Tehuantepec by the Ávila Camacho administration.[11] The Mexican government started to repair the track, housing, warehouses, and ports that the administrations from 1925 to 1938 (the period when Rolland was not head of the free ports) had ignored. The Salina Cruz dry dock's machine shop was in a ruinous state, but even with the budget for improvements squared away, the government was slow to distribute the funding. The sluggish pace of reconstruction chilled locals' expectations about the community regaining its place as one of Mexico's most important ports.[12]

A number of "citizens' committees," labor organizations, and local officeholders—who possessed little interest in World War II—meanwhile were calling for national leaders to end or reduce the influence of the free ports. They argued that the Free Ports Commission's "arbitrary practices" were bad for the town. Rolland rubbed people in Salina Cruz the wrong way. They found him off-putting, and they seem to have little recognized how the free ports were supposed to work or the fact that it was Rolland more than anyone else who had pushed the federal government to revamp the port in the first place. The protesters wrote to the president, to little avail. He had no intention of closing the free ports. Shipping had increased since the dredging of the channels and the outbreak of World War II.[13]

Rancho Santa Margarita

Rolland spent considerable time at the ports but not all of his time. In 1941 he and his wife Rosario Tolentino purchased the

Santa Margarita, a ranch on the outskirts of Córdoba, Veracruz. It was not a huge estate, but it was fairly large, about 247 acres. The buildings on the old ranch were rough. The main house was a large and open building supported by thick colonial walls and square brick columns that held up a large terracotta-shingled roof. To the sides of the house were other buildings, which were run down, but Rolland, along with some of his children and local workers, repaired them to some respectability. One building became a guest house. Other buildings housed the families of a handful of laborers who worked for Rolland. The place looked as it was—constantly under construction.

The scenery was picturesque. From the unpaved road leading up to the ranch you could see the town, mountains, and a vast, rolling greenness. A small river made up one of the property's boundaries. Rolland repaired a modest but beautiful wood-and-metal bridge that crossed the river.[14] He built a pumping station to water his crops, and he constructed tanks to store the water. There was also a stream that flowed only during the rainy season.[15]

Rolland had bought the estate to get away from work and recoup his thoughts, but he couldn't help but turn it into an ambitious enterprise. Coffee plants were already being cultivated on some of the land. He took out loans to purchase fruit-tree saplings. He "covered every meter" of the property, setting out twelve thousand orange trees and twenty thousand banana plants. He made his own pesticides and fertilizers, using materials from Las Minas. He hoped to cover the initial costs quickly and have sufficient success to be able to build credit on futures if he needed it.[16]

Things did not go according to plan. It rained too much, too often. Hurricanes caused serious problems in his orchards. He moved away from bananas, planting instead mangos and more oranges. He also took on some animals—a few horses and some cows. He built pigsties. As time passed he put together some chicken coops, which became home to more and more chickens. He cobblestoned roads on the property. Locals often stole his oranges, prompting him, perhaps using some of those same locals, to build conrete-block walls around the ranch.[17]

Rolland bought the ranch in hopes that his children and a growing number of grandchildren would visit. When not at the ports or in Mexico City, he worked on fixing up the property alongside Tolentino and their daughter Ana María, whom everyone called Anis. She had grown into a young teenager who was extraverted and friendly. As Rolland aged into his sixties, some though by no means all of his hardness began to melt, especially with his family. Now that his children were grown, he yearned to see them and their children more.[18]

The Suchiate Commission

Rolland could never dedicate all of his time to his family, however. The ranch was not the only thing feeding his dreams and headaches. In addition to managing the free ports, Rolland was designated by Ávila Camacho to coordinate a massive project in the southern state of Chiapas, which borders Guatemala. The revolution had hardly touched Chiapas, one of the most economically undeveloped and disconnected states in Mexico. Cárdenas had been the first president to make a strong, serious effort to increase agrarian reform and educational programs in the region.[19] Ávila Camacho built on this foundation. Rolland had for his part become one of the go-to engineers for projects designed to bring the fringes of Mexico more concretely into the Mexican nation. For Rolland the Chiapas endeavor fit well with his goals of extending the free ports and further incorporating Mexico's less-developed peripheries into the Mexican nation.

The Ávila Camacho administration called the project the Comisión del Suchiate, or Suchiate Commission, named for the river that serves as the border between Chiapas and Guatemala. Ávila Camacho decreed the commission into existence on February 14, 1943. It was one of a number of "river basin" commissions that the government initiated during the early 1940s.[20] The goal of the program was to solidify the border, build an international bridge, improve municipal facilities in a number of towns, improve irrigation, reduce flooding, increase the acreage under cultivation, build a new town for workers, build a new port (including a free port), raise statues of prominent figures, construct hospitals, schools, and government offices, install climatological stations, and link them all together and with other parts of Mexico via new roads, telegraph lines, radio stations, railroads, and shipping. All of this would be done to organize and improve the economy for the benefit of the region and the nation, as well as to place Chiapas more firmly under central state control. Hoping to secure his own legacy, Ávila Camacho pushed Rolland to complete the project before the end of the presidential term in 1946. It was a ridiculously ambitious schedule, especially considering Rolland's other responsibilities.[21]

The project involved a number of government ministries, an army of workers, and dozens of planners and managers headed by a cadre of engineers, all ultimately under Rolland. To get the initiative under way, Rolland requested 200,000 pesos to begin river and irrigation works, 250,000 pesos for road construction, and 500,000 pesos to begin a railroad trunk line that would connect the town of Suchiate to the site of the new port, which was given the name Nuevo Morelos. To continue the project, Rolland put forward a budget requesting 60 million pesos, which was to be spread among the projects of the various ministries. The project made some progress with roads, railways, and irrigation, but it never established a free port. Rolland mainly administered.

Problems at the Ports

The ports of Salina Cruz and Coatzacoalcos meanwhile were not running as smoothly as planned. Workers had made some improvements to facilities but not at a pace able to keep up with increased levels of imports, exports, and Tehuantepec-produced oil.[22] In June and July 1943 the arrival of a number of ships from South America exceeded the port of Salina Cruz's capacity, endangering products that could not sit long in the tropical heat.[23] Although Rolland was an influential figure in the designing of the Tehuantepec Railway, the effort had not been put under the aegis of the free ports, whose authority was under the Ministry of Finance; instead Cárdenas had placed it back under the direction of the National Railways, which was within the Ministry of Communications and Public Works. Ávila Camacho wrote to Margarito Ramírez, who was general manager of the National Railways, telling him to end the interruption "of the good functioning of the free port," to quash worker inquietude, and to regain the trust of shipping companies. But letter writing did little to solve the problem.[24] One of the other leading figures of the National Railways, engineer Manuel Buen Abad, served with Rolland on the board of the free ports and the Suchiate Commission, but there remained difficulties about getting the railways, free ports, and Ministry of Marina, which ran the fiscal ports, all on the same page. Rolland and other engineers were overstretched and hampered by infighting and bureaucratic complexity.[25]

Rolland was most annoyed with the Ministry of Marina, headed by Rolland's now former friend, Heriberto Jara. It is unclear when their relationship soured. As port congestion intensified, Jara constantly battled with the free ports, wanting them under his jurisdiction. He made a fuss over a fence built by free port workers in Salina Cruz. He complained bitterly that the free ports were "hogging all they can" and taking space that could be used for regular port operations. It made little difference that Ávila Camacho had sent out a decree to the Ministry of Marina ordering it to provide the free ports with the materials they needed.[26] In opposition to Rolland and Ávila Camacho, Jara had a group of men destroy the contested fence and then impede the construction of a new one.[27] The mayor of Coatzacoalcos meanwhile was upset that he could not use the docks of the free ports as he saw fit.[28]

As if that was not enough to complicate matters, other people were confused and concerned about how the free ports operated. The governor of the state of Veracruz, Adolfo Ruiz Cortines, had no idea how the free ports were supposed to function. He would not receive copies of Rolland's publications on the project until

December 1945. Local officials were dumbfounded about how taxation worked and did not work in the free port.[29] Leaders of a once mighty union, the Confederación Regional Obrera Mexicana (CROM), or Regional Confederation of Mexican Workers, complained to the president that managers of the free ports had fired workers.[30] This rabid infighting and red tape, combined with the increased demands caused by World War II and Rolland's involvement in the Suchiate project, caused a drop in the amount of cargo getting to the capital. Battles between bureaucracies had become a serious problem.

World War II had meanwhile brought the United States, and Mexico with it, into war with Germany and Japan. The Japanese attack on Pearl Harbor on December 7, 1941, forced a more direct entry into the war for the United States. Mexico broke off relations with the Axis powers and then joined the war officially on May 30, 1942, after German submarines sank Mexican oil tankers. The Ávila Camacho administration agreed to allow the U.S. military to enlist Mexicans living in the United States and to establish air bases in Mexico.[31] Mexican officials trained women in nursing, telegraphy, and other means of communication. Able-bodied men ages eighteen to forty-five underwent basic drilling in city squares. Compulsory military service became the law of the land, though not a well-enforced law. The regions of Rolland's greatest interest—the Baja California peninsula, the Gulf, Tehuantepec—became new military zones. The government and businesses struggled to keep up with the rapidly increasing demand for oil and valuable minerals. They pleaded with U.S. officials to better assist Mexico in the production of resources needed for the war. This wartime demand was a boon for raw-resource providers, but it put increased strain on Mexico's transportation network, and it made less likely strong foreign participation in creating industries in the free ports.

The Grand City of Sports

While attempting to quiet the unrest of workers and officials upset about the free ports, Rolland was approached in 1944 by

Neguib Simón, a wealthy Yucatecan sports aficionado of Lebanese descent, about constructing a so-called Ciudad de los Deportes, or City of Sports. Using land on the southern periphery of Mexico City, Simón wanted to build the most impressive stadiums and sports facilities yet imagined in Mexico. Rolland began work on what would become the Plaza de Toros (The Bullring) and the Estadio Olímpico, or Olympic Stadium (not to be confused with the stadium of the same name built later at the Universidad Nacional Autónoma de México). The Plaza de Toros would hold more than forty thousand spectators and the Estadio Olímpico would hold more than thirty thousand people. The Plaza de Toros would be the largest bullfighting ring in the world, and the Estadio Olímpico—which would later become known as the Estadio Azul, or Blue Stadium, in the 1990s because of the soccer team that played there and the creation of a sea of blue seating—would host all sorts of entertainment events, from music concerts to soccer matches. Rolland and his massive crew completed the former on February 5, 1946, and the latter on October 6 of the same year.[32] These huge structures, like Xalapa Stadium, were to bring people together, but it was a commercial endeavor; there was no plan for anything like a garden city.

Ávila Camacho praised the stadium for serving as a model of what his administration was about: progress. To him, it was an opportunity to increase wealth through sporting events. The sports complex would be big and visibly modern. The project also fit with Ávila Camacho's emphasis on physical education, a legacy of the increased attention to physical well-being and hands-on learning that had come out of the John Dewey–influenced revolutionary education programs.

The project was widely celebrated in the Mexico City press for similar reasons. The newspaper *Excélsior* set aside a page to promote it. The newspaper stressed that the City of Sports would "reclaim sports" from lowbrow gamblers and would garner international attention. The sports facilities would bring about "harmonious development" of figures worthy of the Olympics. It would be the "ideal athletic center, not only of our republic, but in the entire continent."³³ The Plaza de Toros in the City of Sports would be "the grandest in the world, exceeding those monuments in Madrid, Seville, and Barcelona."³⁴

The materials used for the Plaza de Toros were considerable. The framing alone used in the construction for the Plaza de Toros consisted of six million cubic feet of wood, built by upwards of thirty-five hundred carpenters. It took six hundred men twenty-one days just to take it down. The completed grandstand stood 136 feet tall. The bleachers covered an area nearing sixty-six thousand square feet. Sculptures of powerful bulls and proud bullfighters stood outside the stadium and adorned the circular walls of the massive ring. With the opportunity to create something grand, Rolland left behind the former progressive-influenced qualms against bullfighting he had professed in Yucatán back in the 1910s.³⁵

The inaugural events were heavily attended and received widespread attention. The first contests in the Plaza de Toros opened in early February 1946 and featured world-renowned bullfighters, including Luis "El Soldado" Castro, the "legendary Spanish matador" Manuel "Manolete" Rodríguez, and Luis Procuna.³⁶ Even though the bullfighting arena was located far from the center of the city, people found their way there to fill the seats. In an example of the easing of government hostility toward Catholic clergy, Archbishop Luis María Martínez gave an opening benediction. The Estadio Olímpico opened on October 6, 1946, with an American-style football game between the Universidad Nacional Autónoma de México (UNAM) and the Colegio Militar.³⁷

Despite the blessings heaped on sports culture by priests and journalists, bullfighting did not always make for the best example of social uplift. On January 20, 1947, a crowd of nearly fifty thousand people rioted in the Plaza de Toros after a match with an "undersized and noncombative bull," the goring of a stadium employee by another bull that jumped the barriers, and all-around poor performances by the bullfighter Lorenzo Garza, who attacked an aficionado with his sword after the fan insulted

Garza's mother. In the "battle of the century" that ensued, people threw seat cushions or lit them on fire, "punched holes in the clock, wrecked the loudspeaker and floodlight system, and roared 'swindle' at everyone involved in the program, including the bull." The radio announcer shut down the broadcast. The damages were estimated at $50,000 (U.S.). To bring an end to the out-of-control situation, "policemen and firemen arrived, the former swinging the butts of their rifles. Both Señor Garza and his tormentor were hauled off to jail, where the former was held incommunicado and the latter was treated by his custodians as a guest of honor. He played poker with them all night." So went the civility, honor, and high morals of sports.[38]

The City of Sports was a reflection of the times. The Ávila Camacho years were the start of an economic boom in Mexico. The industrial sector grew upwards of 10 percent a year.[39] Between 1940 and 1954 overall rates of growth in Mexico averaged above 5 percent.[40] Tourism increased significantly. The government worked closely with business leaders on massive projects. As they saw it, progress was under way. However, the wealth created in the 1940s, as during the Porfirian era before the revolution, disproportionately went into the hands of a small number of elites as well as a slowly growing middle class, further exasperating extremes in economic inequality. Rolland, though still calling for a more progressive Mexico, had lost much of his revolutionary zeal. He still stubbornly espoused his single-tax dreams, but his actions show a desire to build things, to leave a physical legacy, and to increase his wealth.

Sailing on Trains

As World War II wound down and the Cold War between the United States and the Soviet Union expanded, American officials retained their interest in Tehuantepec, arguing that a canal there would still be beneficial to the United States. Although the Panama Canal moved supplies and U.S. naval ships between the oceans, it had reached its capacity of about 30 million tons per year. Disagreements between the U.S. and Panamanian governments over the canal and military bases in Panama further pushed some U.S. Congress members to call for a canal across Tehuantepec. Sen. Dennis Chavez of New Mexico argued that "negotiations should be started immediately with Mexico to bring about the construction of a canal from Salina Cruz to Puerto México . . . based upon a strict understanding that the sovereignty of Mexico would be respected."[41] Rolland, in theory, supported the idea of a Tehuantepec canal, but he did not see it feasibly happening, at least not in the way that Senator Chavez stated, a way that respected Mexican sovereignty. A canal would cost upwards of $13.6 billion (U.S.).[42] The Mexican government couldn't afford to put up even half that amount, which meant that U.S. investors would have to foot the majority of the bill. A canal project mostly funded by Americans did not bode well for Mexican sovereignty.

For a solution Rolland turned to an old idea of "one of the most noted of American engineers," James Eads. Captain Eads, as he was known, had been a well-recognized figure in the mid- to late nineteenth-century United States. A longtime resident of St. Louis, Missouri, Eads had developed diving bells and ships to recover cargo that had fallen into the Mississippi River. He had "devised and furnished the [U.S.] government with its first and most useful armored steamboats." He built "the jetty system for deepening the channel of the Mississippi River," and his grandest vision, still incomplete at the time of his death in 1887, was the Tehuantepec ship railroad.[43]

In the early 1880s Eads had envisioned loading fully laden ships onto trains that engines would pull across a multitrack railway spanning the Isthmus of Tehuantepec. Eads had been driven to action by the French engineer Ferdinand de Lesseps's call to build a sea-level canal across Panama. Eads did not believe that such a project was feasible and that Tehuantepec was a closer and more logical place for a transisthmian route. To Eads, Tehuantepec would serve as an extension of the Mississippi River into the Pacific. The ship railroad, Eads argued, would be dramatically cheaper to build than a canal and would serve the same purpose. The governments of Porfirio Díaz and Manuel González had initially supported the plan. They knew Eads. He had worked in Mexico before, on the Veracruz and Tampico harbors. Eads had surveyed Tehuantepec, and he had a strong track record as a successful engineer. Eads also had a number of wealthy and political backers in the United States, though the U.S. Congress ultimately shot down a proposal to incorporate and subsidize Eads's project. After Eads's death in 1887, the project ultimately fell apart.[44]

Rolland's ship-carrying railway built directly on Eads's vision. The basic concept was the same—move ships from the Atlantic to the Pacific and vice versa using train engines pulling containers set upon ten to twelve tracks. There were, however, differences. Ships had increased in size in the six decades since Eads's death. The system would have to be bigger. Rolland also realized a problem with Eads's original plan. In Eads's designs ships were to be loaded directly onto a large platform that was shaped to fit and grip the ships of the day. It would then be pulled by multiple locomotives. But according to detractors, the design did not address the problem of wear and tear on the ship as a result of the overland transport. Rolland's design placed the seagoing vessels in huge containers filled with water to absorb the shocks of overland travel.[45]

Rolland and other officials associated with the free ports drew up the initial studies for the project during the last year of the Ávila Camacho government as the president transformed the ruling party yet again, this time to the Partido Revolucionario Institucional (PRI), or Institutional Revolutionary Party. The party dropped its military component, and in 1946 Alemán became the first civilian president of postrevolutionary Mexico. The ship railway was one of the first projects submitted to his government.

Alemán built heavily upon the moderating, pro-industrialization, protourism, and pro–middle-class policies established by his presidential predecessors, especially in the years following the 1938 oil expropriation. Alemán also represented a new and younger generation of leaders, a generation that had not fought in the revolution but instead were the college-educated sons of revolutionaries.[46] They brought the trends of technocratic leadership, civilian rule, and state-sponsored economic growth to new heights. Detractors criticized them for leaving behind the poor and abandoning revolutionary redistribution while living corrupt, playboy lifestyles.[47]

Alemán initiated his own façade of an anticorruption campaign when he first entered office—it had become a tradition of sorts.[48] This time Rolland was not a target. Instead Rolland remained an important figure in infrastructural development and trade. There are no existing documents in which Rolland expressed his personal feelings about Alemán. Rolland was definitely not an intimate member of Alemán's inner circle of close generational friends, but he appreciated Alemán's probusiness leanings, modernizing impetus, and focus on urbanization.

In their initial reports Rolland and his close associates provided meticulous details about what the ship railway would entail. In publications meant to influence the Mexican government and entice U.S. investment, Rolland laid out the numbers on the required amount of track and the necessary rails, ties, filling, cuts, grading, and bridges large and small. He went over plans for a power plant and the need for additional studies on the terminal ports, sidetracks, and the dry docks. He discussed the issues of building in the hills in the center of Tehuantepec and the lowlands of the Gulf and Pacific coasts. Each dry-dock carriage, which could back into the ocean in order to accommodate the ships, would be connected to a series of waterproof 22,000-horsepower electric engines. Each dock would be remotely controlled by radio until connected to locomotives that would carry the ships across the 185-mile long isthmus at 25 miles per hour.[49]

Rolland, arguing that red tape slowed progress, pushed for the ship railway to fall under the jurisdiction of the free ports. It was officials of the free ports after all who had conducted all of the studies. And although there would be a fee, having the ship service operate according to the guidelines of the free ports would allow duty-free movement of goods, which Rolland contended would entice global capitalists and build goodwill between Mexico and its trading partners. At the same time, the new service, along with a highway slotted for construction across the isthmus, would continue to improve Mexico's east-to-west infrastructural development. Alemán would ultimately provide the free ports with broader jurisdiction, while putting them and most Tehuantepec operations under the supervision of the Ministry of Finance.[50]

Rolland and his new "land bridge" garnered widespread attention. The *Blytheville Courier News* of Arkansas declared, "Rolland is no amateur at converting large-scale projects into realties. He is the man who introduced reinforced-concrete construction to Mexico. He is also responsible for some of the best buildings, ranging from luxury hotels for tourists, one of the continent's most lavish gambling casinos (now out of existence), the Jalapa stadium, port installations, and Mexico's recently completed Sports City."[51] The *New York Times* called the ship railway a practical solution to an old problem. Rolland accompanied navy and army officials from Mexico and the United States on tours of the isthmus during which he discussed his ideas for the ship railway. He happily showed off models he built to display his desired outcome for the project.[52]

The plan fit nicely within Alemán's push for infrastructural advancement and a closer relationship between the governments of the United States and Mexico. The mileage of roads in Mexico had expanded significantly. "Mexican motoring" had become a fad for adventurous American tourists.[53] The Alemán administration obtained an $8 million loan from the Bank of America and the Mercantile Commerce Bank & Trust—both U.S. banks—to finish a highway across Tehuantepec.[54]

Over the course of the remaining years of the 1940s Rolland continued to compile studies and to promote his grand idea. He had made some modifications to it by 1949. Rolland changed the estimated speed of travel to 20 mph instead of 25 mph. The power required to move the dry-dock tank cars was upped to 30,000 horsepower. Modifications were made to allow for ships as large as thirty-five thousand tons, which would include all "but half a dozen passenger liners and the largest battleships and aircraft carriers" then on the seas.[55] Rolland also told a crowd in Los Angeles that "much of the space in the bottom of the tank car, between its square corners and the ship's rounded hull, would be filled by 'air bags,' which would reduce by 50% the weight of water that would have to be carried to float the ship within the tank car."[56] He constantly worked to improve the design and his sales pitch.

Not everyone in the United States, however, was sold on his idea. Rolland traveled to Washington DC in 1949 to promote the ship railway. Much to his chagrin, some U.S. engineers contended "the enormous weights, pressures, and tensions involved were inconceivable."[57] The management of the private companies that operated the two free ports in the United States—Staten Island and New Orleans—had campaigned against the free ports in Mexico, which they saw as competition.[58] It was a serious blow to his fundraising tour. As the decade came to end, Rolland continued to push the ship railway while working to expand development in the free ports and on the frontiers. Retaining the support of his superiors, he persisted despite frustrations.

The World's First Stationary Dredge

In Salina Cruz nature and unions would do their best to frustrate Rolland further. Waves continued to bring in massive amounts of sand, often clogging the navigation channel and costing the Mexican government millions of dollars to dredge the channel with special boats. These dredges were operated by a local union, which, according to Rolland, too often shirked its duties, "besmirching" the reputation of the port among businesses that operated large ships.[59]

Needing a solution, Rolland and the other members of the Free Ports Commission drew up plans for the "world's first" stationary dredge, an idea Rolland had first developed in 1944. It was to consist of a massive block structure with mechanical dredges connected to it and six massive suction pumps that would suck up sand and then pump it through a pipe sixty-six hundred feet long. The sand would be ejected outside the port and then carried toward the southernmost state of Chiapas. In addition to the dredging arms, suction pumps, and sand-carrying pipe, the stationary dredge consisted of a number of large motors, generators, electric winches, and a large switchboard. There was an almost identical "substation" that helped propel the sand out of the harbor. The engines and the switchboard came from the United States, but most of the dredge machinery came from Werf Conrad, a Dutch company that specialized in dredges and that also sent specialists to help set up the dredge. Rolland further double-checked blueprints with U.S. engineers.[60]

The prep work to get the stationary dredge fully operational was in and of itself arduous. The base structure, essentially a massive concrete block, had to be sunk and settled. On top of this massive block, construction workers built a 197-foot-long "enclosed space to install standard dredging equipment." The suction pumps had to be sunk deep into the sand and sea. This was no easy task: "This entailed a strenuous operation because each time the opening was made, sand rushed in [in] such quantities that the lagoon was filled up and the pipes stopped functioning. After several trials and arduous work, we finally had to construct passageways of reinforced concrete sheetpiling, in 23 and 32 ft. lengths, in front of the pipes, through which a drag scraper operates pulled by a winch and guided by steel cables passing over a pulley installed on a float anchored at sea."[61] This operation was just the start of a larger process of removing the vast amount of accumulated sand, since the stationary dredge's suction pumps were designed to handle only the amount of sand carried by the waves. An entire section of the Salina Cruz coastline would have to be changed for the dredge and sand-bypassing plant to properly function.

Both the ship railway and stationary dredge, despite all the time spent on them, were still incomplete as the 1952 presidential election approached. Engineers and construction workers had made significant headway on the stationary dredge, building the main structure and attempting to remove sand and change the coastline to make the dredge and its suction pipes functional. The ship railway, however, had run into a serious roadblock and remained in a preliminary phase. Rolland had failed to obtain the necessary private capital to move ahead with creating the ship containers, railroads, and all the associated equipment. Rolland also faced mounting criticism over both projects in terms of feasibility and costs, and local unions were opposed because their members feared losing their jobs.

During the Ávila Camacho and Alemán years, Rolland had undertaken some of his grandest projects. He silenced criticism over the viability of the City of Sports and went ahead with designing and presiding over the creation of the largest bullfighting ring in the world and one of the most impressive sports complexes in the Americas. Rolland had expanded his oversight of infrastructural projects into Chiapas and the Guatemalan border region. He faced more difficulties with his railway and port projects, which were crucial to his longtime development schemes for Mexico. But they appeared to be moving forward, albeit very slowly. It was unclear, however, whether the next president would keep Rolland, now in his seventies, in his position as general manager of the free ports. It wasn't clear if the new administration would continue to back his projects at all.

12

Out of the Ports and into the Hills

As the 1952 presidential election filled city streets with the typical campaign flyers, many of Modesto Rolland's projects and partners placed him in an increasingly probusiness and politically conservative light. He encouraged free trade even as Mexican free traders found themselves in the minority.[1] He advocated for foreign investment while mingling with financiers and government officials from around the world. He denounced strikes and labor unions, and he had become a staunch opponent of Mexico's agrarian reform policies. Rolland had not amassed a fortune, unlike the greediest of government and business leaders, but he lived comfortably. He had his longtime residence in Mexico City and the ranch outside Córdoba, Veracruz.

The prodevelopment administrations of Manuel Ávila Camacho and Miguel Alemán had more or less meshed with Rolland's own goals, except for the high level of industrial protectionism. Rolland had lost his place as a government undersecretary, and none of the presidents who served in the 1940s took seriously his resilient Georgist and Progressive-era prescriptions, but both administrations had pushed a pro-middle-class, moderate, and industrializing agenda that worked with Rolland's longtime goals of uniting the nation though infrastructure and connecting Mexico with the rest of the world through trade.

The early 1950s would prove to be the last years of Rolland's career. A few of his colleagues from the revolution remained in prominent circles, but a new generation of technocrats and entrepreneurs had begun to fill government bureaucracies and the business offices of downtown Mexico City. Rolland continued to lay out his case for a better Mexico, but he had made a number of enemies over the years, a reality that would catch up with him. Ultimately Rolland was pushed into retirement at his ranch outside of Córdoba, Veracruz, and into a pleasant cliché. He gave up all professional pursuits outside of his ranch operations, instead spending time riding his horse, caring for his orchards and animals, and attempting to make up for his past faults as a father and husband by being a dutiful spouse and more attentive father and grandfather.

Overall Rolland had been a relatively successful engineer, especially considering his circumstances, but what stands out most about the end of Rolland's professional career, and to some extent his entire career, is that a number of the projects he had invested so heavily in—the ship railway and the stationary dredge—failed. The free ports continued but with an uncertain future and with results that were far from obviously positive. How these projects met their demise and what we can learn from their failure (and limited successes) are the subjects of this chapter.

How to Upset an Incoming President

The beginning of the end for Rolland began as it had for many of his colleagues—with a presidential election. During the 1952 election the PRI candidate, Adolfo Ruiz Cortines, put forward a mostly rhetorical question to Mexico: "What is the best way to combat misery and communism?" Rarely missing a chance to give his opinion on matters of moving Mexico forward, Rolland replied publicly just months before the election. His response was filled with arguments familiar to anyone who knew him, and they marked his return to public political commentary.

Rolland titled his double-columned, eleven-page response "Efectiva manera de evitar la miseria pública y combatir al comunismo" (Effective way to avoid public misery and combat communism). He published it in two popular trade journals, *Revista Industrial* and *Construcción Moderna*. In this essay Rolland acknowledged the presidential candidate's probable good intentions and the importance of the technical issues that he raised. In Rolland's view, however, that list of issues was far from complete. Ruiz Cortines also needed to address education, hydraulic resources, irrigation, ports, municipal administration, the armed forces, public health, and fertilizers. Beyond that point Rolland was less sympathetic. He called Ruiz Cortines's questions a "curtain of smoke" and nothing but trendy topics that failed to address seriously Mexico's basic underlying problems.[2]

In the subsequent pages of his response, Rolland yet again aired his frustrations with the revolution. Rolland wrote that the revolution had resulted in a number of "national disasters." He blamed politicians for continuing to use the agrarian problem for political advancement instead of tackling it with technical expertise, leaving the "agrarian problem unresolved." Rolland then listed other fundamental issues: the exploitation of labor organizations by the powerful under a "dictatorship" of corrupt labor bosses—practices that failed to benefit workers or resolve production problems; the continuance of dysfunctional municipal governments plagued by a political tributary system; a lack of effective and honest elective processes, which stymied democracy and failed to put the most skilled people into positions of power; and a lingering problem with the Catholic Church, especially its ownership of a vast number of properties.[3] Rolland had been making most of these arguments since the 1910s. They were throwbacks to the Progressive era. But even if his thoughts came off as passé, many of them still retained some truth. Politicians had used social policies as tools for personal aggrandizement and political expediency instead of as a genuine means to help people. There were serious, long-lingering problems of corruption in labor organizations. Government at all levels had failed to end tributary practices and to enact basic democratic practices.

His attack on Church landholdings, however, was even more erroneous than when he first stated it. The Church did not own anything close to the massive number of properties it had possessed in colonial times. It had lost a huge portion of its wealth, even if it still had a strong grip on Mexican culture. Rolland himself had not fully escaped Catholicism. Rosario Tolentino remained a devout Catholic. Rolland begrudgingly went with her to mass occasionally in order to appease her. Most of Rolland's children and grandchildren remained at least nominally Catholic.

It is also surprising how much Rolland continued to harp on the *ejido*. For the previous decade Mexican presidents and their ministers had slowed the ejido program to a near standstill. They had instead focused more on large-scale commercial agriculture, the use of new fertilizers, and major irrigation projects.[4] Ejidos existed in great numbers, but ejido populism had died down in the 1940s. Perhaps he was anticipating Ruiz Cortines's move to grant more ejidos during his presidency.

Showing his ever-resilient belief in economic Georgism, Rolland argued that the tax system remained a complicated mess that punished the poor and those people who strove to improve their lands through agriculture, construction, and enterprise but who possessed limited capital. To those officials who knew Rolland well, it must have seemed like he was beating a dead horse. No president was going to enact a single tax on land. The administrations of Ávila Camacho and Alemán had increased government funding by expanding federal taxes such as the income tax, though businesses and wealthy members of society often got away with paying little in the way of taxes, a problem that continued to hurt government coffers while exacerbating inequality. Small business owners were harassed with fees and sometimes by people demanding bribes. But in general people were paying relatively little in taxes. By the end of the 1960s Mexico collected less in taxes per person than any other Latin American country.[5] The government's failure to effectively collect taxes has remained one of the greatest weaknesses of the Mexican state, hindering its ability to pay for large and costly public works projects. There was an argument to be made for a single tax on land, even though the vast majority of economists in the 1950s ignored it. It has had many proponents, including the writer Leo Tolstoy, physicist Albert Einstein, philosopher John Dewey, and, more recently, Nobel Prize–winning economist Joseph Stiglitz. It remains a debated idea that has sparked global interest if limited enactment. But few in the Mexican government supported it. Most economic advisors in the 1940s and 1950s pushed for a progressive tax system that more closely resembled those in other industrialized nations but with strong tax breaks for certain industries and equipment and protective tariffs on a large number of foreign manufactured goods.[6] And the few avid Georgists in Mexico, among whom Rolland was the most prominent, never laid out a precise plan for how to move from indirect taxes to a land-value tax.[7]

The public letter to Ruiz Cortines did Rolland no good. It showed a man passionate about the welfare of his country but also a man clinging to long-held convictions of which he was unwilling to let go, even when they were no longer accurate or attainable. Rolland's antiunion stance did not make him any allies within labor organizations, especially those who worked the ports and railroads. Calling politicians corrupt and the revolution a failure did not win him many friends in political circles. He was not a party stalwart. He irritated people on both the political left and right.

It did not help his case that the free ports had not proved particularly successful. They had faced all sorts of difficulties, including World War II demands (and lack of demand for foreign industrial workshops), changing government policies, inconsistencies in funding, wavering support and outright resistance from U.S. financiers and engineers, attacks from other Mexican government agencies, and the never-ending process of keeping Salina Cruz in working order. Although World War II and the subsequent Cold War provided presidential administrations with a number of justifications to crack down on worker radicalism, strikes during the 1940s and 1950s still hampered port and railroad efficiency. In addition to unions demanding better wages and working conditions, as well as the dismantling of certain machines, in particular Rolland's stationary dredge, a protest by multiple sectors of Oaxacan society against the state's governor, Manuel Mayoral Heredia, closed down the port operations in Salina Cruz in late March 1952.[8] Disturbances at the Tehuantepec ports and the railroad that connected them remained fairly common. But excuses only encouraged detractors. Rolland and other officials working in the ports were successful in at least keeping the ports and railway functional. Still, for all the money spent, there was little improvement in the lives of most Tehuantepec residents, limited industrialization, and no astounding growth in Mexico's international trade. The region never developed as fast or as beneficially as the promoters of the free ports had hoped.

Nor was Rolland the only person critical of the government, though other critics provided different reasons for their dissatisfaction. A number of Mexican scholars, including Daniel Cosío Villegas and Jesús Silva Herzog, had essentially called the revolution dead, with the government paying it mere lip service. Cosío Villegas lamented that what passed as progress in Mexico was a sham because the economic and industrialization schemes promoted under Alemán had benefited only a minority of Mexico's population: financiers, the small if growing middle class, urban laborers, and political leaders. Although most critical commentators saw the move away from revolution-era policies as a betrayal, not everyone was so negative. The sociologist José Iturriaga argued that the revolution may have been dead or dying, but that its passing was okay. The revolution had brought more success than failure. A new generation was taking over, and Mexican society had to evolve. He believed that "the people's eternal instinct for progress" would "not have to die with the extinguishing of the Revolution."[9] For Rolland it was not a matter of the revolution coming to an end but of it failing to solve Mexico's basic problems to begin with. It was a sentiment he shared with other technocrats of earlier generations, such as Luis Cabrera, Manuel Gómez Morín, and many of the living members of the class of engineers that had been with him at the outbreak of the revolution.

Condemning Rolland

At the local level in most small towns across the country people's opinions about the success or failure of the revolution, or about the PRI, usually had more to do with local officeholders. Farmers and villagers were more often than not casualties of (and participants in) the political tributary system that Rolland consistently bemoaned. In the words of Sydney Gruson, a *New York Times* journalist who toured the Mexican countryside and talked to people about the upcoming election, "Local bosses, . . . once installed by the national political machine, rule with few checks."[10] Most residents gave little credence to the election process, considering, with justification, the outcome to have been determined well before the actual election. Political bosses organized demonstrations of large union blocs. Most ordinary farmers were ignored or bribed to vote for certain candidates. Social reforms and modernization projects had progressed slowly in rural areas.

Many people saw Ruiz Cortines as boring and stern, a counter to Alemán's excess. The government under Ruiz Cortines would build on the modernizing drive of its predecessors and finally grant women full suffrage, thirty-seven years after Rolland had helped organize the 1916 women's congress in Yucatán. During his inauguration, Ruiz Cortines slighted the Alemán administration, calling it plagued by corruption. The new president then declared that he would be "inflexible with public servants who stray from honesty and decency."[11] Of course by this point this type of speech was almost required of new presidents. Many people, however, took Ruiz Cortines more seriously. He did not tolerate political competition with the PRI, but he came off to many people as honest. He was prudent with his money, and he promised the same with the nation's treasury. He was admittedly probusiness, but he cared little for projects he considered wasteful and quickly built a reputation for his "Spartan morality."[12]

In a letter written by the leaders of a number of unions in Salina Cruz who came together to condemn the managers of the free ports, Ruiz Cortines found his reason to force Rolland's resignation. The letter writers included members of the Salt Workers Cooperative of Marques, railway workers from the Union of Workers of the Mexican Republic, the Union of Longshoremen and Laborers, the Regional Federation of Workers and Farmers, and the Petroleum Workers of the Mexican Republic. They specifically called out former president Alemán, former secretary of finance Ramón Beteta, engineer "Rolant" (Rolland's name was misspelled throughout the letter), another engineer with the last name Macedo, and a so-called *turco* business operator named José Abrahan (perhaps Abraham?) for corruption.[13] The letter writers' goal was the removal of the top free ports officials.

The union leaders laid out a number of complaints, playing specifically to Ruiz Cortines's publicly stated distastes. Macedo, they argued, was almost always drunk. Sex workers and cantinas paid him off so they could operate. Abrahan ran a number of warehouses where he ran an illegal operation that stored tons of corn, beans, and other foods while the poor died of hunger. According to the letter writers, the called-out men had all profited from contraband, using boats and airfields to bring in goods without paying taxes. The supposedly illicit goods were said to include a hundred barrels of American whiskey, of which half went to Beteta and the other half went to Rolland and Macedo, who drank wildly and gambled. The union leaders also argued that Rolland was unapproachable and sold gasoline under the table to certain ship owners. They also accused Rolland of reporting that hundreds of laborers were working for him even though in reality he had a much smaller number of skilled workers on his payroll. Rolland, whom they stated owned houses throughout Mexico, was said to have pocketed the salaries of these imaginary workers. Meanwhile locals who fished in the area and peddlers could not sell their goods in the free ports, the union workers made meager wages, and the city made no money from taxes or the goods in the ports.[14]

For Rolland's opponents the greatest symbol of the corruption in the free ports was the stationary dredge. In Spanish the name of the dredge was the *draga fija*, literally translated as fixed dredge. But according to the union heads everyone in Salina Cruz called it the *droga fija*, or stationary drug, because instead of sucking sand it siphoned millions of pesos from the country and placed the money into the pockets of Rolland and his collaborators. Locals had other names for the dredge as well: the white elephant and the mansion of Salina Cruz. Union members complained that Rolland had spent 20 million pesos of the government's money on the dredge and that there was nothing to show for it except for a strange, massive structure.[15]

The limited evidence that exists contradicts many of the union leaders' accusations. For example, according to people closest to Rolland, he was not the gambling type. He had constructed a building used for gambling back in the late 1920s, but he despised gambling and, usually, gamblers. Rolland almost never drank. By most accounts he was a teetotaler. Nothing suggests he would be illegally importing whiskey, drinking it, and gambling. Illegal alcohol production and sales using free-trade zones had been an ongoing concern among critics since the Porfirian administration created free-trade zones along the U.S. border. One interesting detail is that two months before the union leaders' accusations there was a scandal about scotch whiskey production on the other side of Tehuantepec in the free port in Coatzacoalcos. Producers of authentic Scotch whisky (who capitalize Scotch and use no *e* in whisky) complained through British representatives about Charles Klor, an American from San Francisco who had set up a manufacturing shop of sorts. He had been mixing "scotch malt and Mexican spirits . . . bottled and labeled as scotch and then sold in Mexico as though it had been brought from abroad."[16] This was not the first time that repackaged hooch was recreated in a free port and sold, though past complaints had focused on operations in the U.S. free port at Staten Island. Rolland surely knew of Klor, though it is unclear how much Rolland knew about the specifics of Klor's operation. It also wasn't clear if there was anything illegal about it. But Rolland definitely would not have been interested in drinking whiskey, legal or illegal.

Other denunciations were erroneous as well. Rolland owned two homes and surely stayed in a provided residence while in Salina Cruz, but he did not own properties all across Mexico. In Rolland's publications about the stationary dredge he specifically stated the costs, indicating that one of the goals of the dredge was to reduce the number of laborers and the expenses involved in dredging. Indeed he sent regular reports to colleagues about the projects going on in the free ports and about their costs.[17] Engineers and construction workers had completed the nuts and bolts of the stationary dredge itself, but they were still working on getting it fully operational. Laborers were shaping the coastline and removing sand so more water could enter and the pumps would not get clogged with too much sand. The *draga fija* was expensive to build, time consuming, still not functional, and the butt of jokes, but the stated price tag that union leaders used as an attack against Rolland was close to the estimate he had provided to his superiors and to the public. He genuinely believed in the project, but Rolland and the port workers did not believe in each other. The stationary dredge jeopardized the job security of certain port workers, and Rolland knew and welcomed that fact; he had a low opinion of many of them. When he was with his family members Rolland referred to many of the port workers as lazy and ignorant drunkards. The union leaders had their own agenda. They planned on using the transition of governments to get rid of Rolland, someone they rightly considered an enemy.

The letter writers also appear not to have understood the logic and laws behind the free ports. They noted more than once that the local municipality did not benefit from taxes. They also complained that those people who stored materials in the warehouses did not pay taxes. Of course this was exactly how the free ports were designed to operate. Many residents believed that the free ports' managers were breaking laws. What locals didn't realize was that it was the laws themselves that they did not like.

More than anything else, accusations leveled by union leaders in 1953, as well as complaints from other Tehuantepec residents in later decades about similar development projects, appear to have come from frustration about being excluded and forced to deal with changes they had not desired.[18] They too often found themselves with little real power to influence national policies, and no one in power appears to have asked for their opinions or placed any value on their way of life. Many people in Tehuantepec, including thousands of non-Spanish-speaking Mixe, Huave, and Zapotec residents, had no idea about the free ports or the specific industrial designs for the isthmus. In Salina Cruz most workers were aware of the free ports and the stationary dredge, though they knew little about how they were supposed to function economically or the intricacies of the laws governing the project. Many port laborers and Salina Cruz vendors believed that the free ports jeopardized their livelihoods more than they strengthened them. They were never sufficiently reassured that they would obtain jobs or retain them. They were paid low wages. Those who engaged in fishing were never asked if the projects jeopardized their way of life or if they wanted the region to become an industrial center. And many people were indeed hungry while wealth, often in the form of duty-free goods, flowed through the ports, even when the ports were not operating at their most efficient capacity. Fenced off from the rest of the community, the free port seemed all the more mysterious, isolated, and foreign. It became a symbol of a cold world built with machines that funneled wealth to outsiders.

Even if the worst allegations of corruption against Rolland were not true, his designs for making the region prosperous had largely been a failure. The lives of the vast majority of workers did not improve. Foreign and domestic financiers and industrialists never developed Tehuantepec along the lines he imagined. There were new roads, oil pipelines, and gas lines, which helped fuel international trade, but they also left a heavy environmental footprint. Like so many other engineers who carried out top-down state and corporate designs, Rolland never understood the crucial importance of working closely with local people and within their cultures, of the need to understand and appreciate them. This was one of Rolland's greatest weaknesses. Rolland wanted to improve Mexico according to his vision and in spite of other people. For that, people often resented him and fought against him.

Ruiz Cortines was less than enthused about Rolland before the accusations ever reached his desk. While governor of Veracruz (1944–48), Ruiz Cortines had not been excited about the federal government's free port in Coatzacoalcos. He remembered waiting to obtain slow-coming information about it, the infighting among the different port authorities, and the clogging of the ports with heavy traffic in 1943 and 1944. Ruiz Cortines used the letter from the Salina Cruz union leaders as a way to appear hard on graft and waste. Ruiz Cortines was happy to have an excuse for getting rid of Rolland. Rolland attempted to meet with the president on a number of occasions to explain himself and his plans and to plead for his stationary dredge. Rolland's requests appear to have gone unanswered.[19]

Rolland was replaced as the manager of the free ports in 1953 by another engineer, Antonio Pailles, who served until 1957. He oversaw the dismantling of the stationary dredge, which was sold as scrap metal while the most valuable parts were salvaged for other projects. Before the end of the Ruiz Cortines administration, the president replaced Pailles with Jacinto B. Treviño, a former general and military engineer. Pailles was also a onetime Carrancista who had been forced out of Mexico for joining the Escobarista Rebellion in 1929; he was allowed to return in 1941. The government also brought in a number of other ministry officials to join in the decision-making, complicating things further. The port continued to suffer from the wave-driven sands as technical improvement stalled.[20]

Ruiz Cortines's ousting of Rolland from his managerial position in the free ports was a major blow to the aging engineer's career and his dreams for Mexican development, but it did not completely end his professional career. In 1954 Carlos Lazo Barreiro, an architect serving as secretary of communications and public works, invited Rolland to assist with the preliminary master plans for a new building complex for the ministry he led. It was a notable system of buildings drawn up to house the growing bureaucracy. Most impressive was a massive mosaic created on the tallest building by none other than Juan O'Gorman, the architect and artist who had created the mural at the Mexico City airport that Rolland had ordered changed. O'Gorman had also worked with Lazo to create a now famous mosaic on the walls of the newly built Central Library of the Universidad Nacional Autónoma de Mexico, which he had designed in the same style that he would bring to the new communications building. The mosaic on the communications building celebrated Mexico's technological and infrastructural advancements—advancements that Rolland had helped create. It doesn't appear that Rolland and O'Gorman worked together, as each was participating in different stages of the project.[21] Rolland's only subsequent professional presence was during the next presidential election. In 1958 he accompanied the PRI candidate, Adolfo López Mateos, on a tour of the Baja California peninsula.[22]

Retirement

After his ouster from the free ports Rolland spent most of his days at his Santa Margarita ranch.[23] He sold his Mexico City house sometime in the mid-1950s, thus making the ranch his permanent home. He worked among his plants and animals. Following the Cuban Revolution of 1959, Rolland switched from growing oranges to sugarcane because the U.S. boycott of Cuban sugar made the sweet product a more profitable commodity. Rolland invested in a small sugar mill to turn the juices of the cane into *piloncillo*, which is the everyday brown sugar, usually sold in cone-shaped units, that is common throughout Mexico. Rolland wrote to his children and grandchildren, occasionally hosting them. His children, all well into adulthood and undertaking careers of their own, helped with projects alongside the people Rolland hired to work the ranch. Anis had become a chemist and had married another chemist, a man named Jorge Salcido. The grandkids played in the river. For holidays and other special events the family took a group photo, something they had done with some regularity throughout the 1950s.

It is tempting to interpret these photos. Rolland is always front and center, the proud patriarch of a large, transnational family. In a 1959 photo the family sits under a massive painting of Saint Christopher, who is holding baby Jesus. Besides being a patron saint of travelers, Saint Christopher has historically been portrayed in stories and art as carrying an extraordinarily heavy baby Jesus across a river—a heavy faith upon his back. Rolland, surrounded by all his progeny, is lined up directly below Christopher, reinforcing Rolland's importance in savior-like fashion, as if he were a part of some divine chain of being, the burden bearer of his family. Rolland's faith, like Christopher's faith, had been tested. Rolland's faith, however, was a faith in progress, to be a carried out by a modern and secular people focused on planet Earth, not metaphysical fancy.

In an earlier family photo, from 1952, taken at Rolland's previous residence in Mexico City not long before he sold the house, the wall that makes up the backdrop is covered with black-and-white framed and matted photos. They are all of Xalapa Stadium. The picturesque structure made a natural choice for wall art, but it is interesting that it was what Rolland emphasized most. The photos also showed the highlands of Veracruz, which he loved. The stadium and the City of Sports, which also appeared in many of his mounted photos, were also his most visible legacy. They were tangible, solid, and long lasting. He was proud of them, and he wanted them and himself to be remembered. There were no photos of the free ports or of his times as a propagandist or land reformer. The products of his greatest ideals, they left a legacy more difficult to grasp and with mixed results. They were relegated to archives and stories.

As the 1960s got under way, Rolland tended to his sugarcane and pigsties while President López Mateos fought migraines and railroad workers. The PRI struggled to maintain a façade of revolutionary convictions in the face of the new torchbearer of Latin American revolution, Fidel Castro and the Cuban Revolution. Critics of the PRI and López Mateos protested the increasingly obvious single-party "dictatorship" of "Boom and Progress, of Technology and Democracy."[24] Fifty years after Francisco Madero's call to arms, the Mexican state and economy appeared to be as much of an expanding and problematic continuation of Porfirio Díaz's "order and progress" as a new radical society based on equity and social justice. López Mateos attempted to push the PRI back in a "revolutionary" direction, nationalizing electric companies and once again increasing the number of ejido grants. And although Porfirian legacies lingered, the revolution had brought about significant changes: regained pride for many ejido farmers, an increased dialogue between hundreds of thousands of Mexican workers and the government, much more infrastructure, and an expanded state. But the PRI had a strong grip on power. Presidents changed, but there was no real democracy. The leaders who had taken over the party since Alemán were mostly technocrats—university-educated civilians interested in technological advancement, capitalist enterprise (often driven by the state), and urban industrial growth. Mexico City grew by leaps and bounds. By 1960 Mexico had become a majority urban country. But along with urbanization grew slums and an inequality that seemed immune to change.[25]

It is impossible to surmise exactly what Rolland thought about his legacy. He reflected on it occasionally, but mostly he kept busy running Santa Margarita. He had begun to make money selling sugar, and along with Tolentino he managed a number of laborers who worked the ranch: an overseer, a stable keeper, and another young man who fed some cows that Rolland had acquired. The family of a gateman lived in a small wooden house at the entrance to the ranch. Tending the pigsties required a couple of workers and the occasional veterinarian visit. Workers fed the sugar mill with cane harvested from Santa Margarita and other farms. Other laborers produced the sugar cones by heating the cane juice, cooling it, pouring it into molds, and packing it. Rolland and other workers cleared land with a tractor. Rolland and Tolentino also spent considerable time expanding their flock of chickens. Rolland had huge coops constructed, and by the early 1960s he and Tolentino had approximately ten thousand chickens, mostly hens. Tolentino tended to them and operated an egg business. When the grandchildren visited, Tolentino had them help collect eggs and then sort them by size using a sorting machine. Finally turning a profit, Rolland kept busy overseeing daily operations, sales, and debts.[26]

He did not escape thoughts about his past completely. In discussing his time as a young Anti-Reelectionist with his grandson Jorge, Rolland talked about how police loyal to Victoriano Huerta had threatened to kill him. When he could not sleep, Rolland would sometimes rework past ideas, especially calculations for the ship railway. It ate at him that the project never came to fruition. Despite all attempts to find peace, bitterness occasionally surfaced about the workers who fought him when he was manager of the free ports. He rarely discussed regrets, though he told family members that he had gained many enemies during what he perceived to be an honest, long-fought battle to improve Mexico.[27]

Rolland did write a memoir of sorts: "Vida y hechos del Ing. M. C. Rolland" (Life and works of engineer M. C. Rolland). Following discussions with his grandson Jorge, who was training to become an engineer, Rolland scrawled out six pages of his hurried autobiography in sloppy cursive handwriting. It must have taken all of a few minutes. In the first two pages he wrote proudly of his work as a young man at the National School of Engineering. Rolland also provided a paragraph listing of other accomplishments: introducing reinforced concrete to Mexico, his building of houses, the construction of the petroleum terminal in Progreso, the Hotel Chula Vista in Cuernavaca, and the Foreign Club on the outskirts of Mexico City. He mentioned the construction of Xalapa Stadium and the docks and warehouses in the free ports. He spent considerable time discussing the City of Sports, especially the Plaza de Toros. As shown in the photos on the walls of Rolland's home, it was ultimately his physical legacy that he cherished most, his tangible accomplishments. He did briefly discuss his time working in politics and government circles, particularly mentioning his time under Venustiano Carranza and Lázaro Cárdenas. But perhaps most telling of his frustration with the politics of the revolution is Rolland's own statement about retirement on the ranch: "Perhaps I am more useful here than as a political engineer; at least there is peace of mind."[28]

Rolland had been suffering complications from diabetes, which were getting worse. Despite his lifelong promotion of science, he treated his diabetes with homeopathic medicines, a common practice throughout much of Mexico. He was also fond of sweets. He commonly sucked on low-sugar candies, but his youngest grandchildren, the children of Anis, would sneak him regular candies, which he gladly devoured despite Tolentino's grumblings. Eventually the disease and age took their toll.[29]

Rolland died in a small hospital in Córdoba on May 17, 1965. The family buried him in the Panteón Jardín cemetery in Mexico City, the resting place of a number of Mexico's most famous political and cultural icons, including the actor Pedro Infante, soccer player José Miguel Noguera, and sculptor Mardonio Magaña. Rolland's funeral was a solemn affair. As the casket was lowered and caretakers began covering it with earth, one man, a professor named Eliseo Bandala Fernández, decided to speak. Bandala was the husband of Tolentino's sister Ana and one of Mexico's most prominent educational leaders. Like Rolland, he had worked in his own way to construct a modern, more unified nation after the revolution. Sharing similar interests and long careers as revolutionary officials, Bandala perhaps understood the significance of Rolland's work better than most of the people in attendance. Bandala praised Rolland as a great man of many talents. Bandala told the family that they should be proud of how much Rolland had influenced the course of Mexican history and how much he had achieved for himself and his family. Within months Rolland's sister Victoria, the sister who had suffered from a debilitating mental illness and lived with Rolland until he moved to the ranch, died. She was buried alongside Modesto.[30]

Rolland had lived a storied life that included an impressive half-century of service. Born in the peripheral-yet-global town of La Paz, he had beaten the odds and had risen to become a university-trained engineer who worked with some of Mexico's leading minds and political figures. He helped ignite and then survived a long and bloody revolution, continuing to thrive, finding himself—through a combination of skill and luck—on the right side of subsequent conflicts.

One thing that Rolland's life shows is that technocrats have been influential in government circles within Mexico since before the revolution. What many critics of the 1960s PRI failed to acknowledge was that the Mexican government had been to some extent run by technocrats of "Boom and Progress" since the Díaz era. Mid-twentieth-century technocrats were intellectual descendants of people like Díaz's close advisors José Y. Limantour and Justo Sierra, as well as engineers such as Francisco Bulnes and Rolland's mentors—Manuel Marroquín y Rivera, Miguel Ángel de Quevedo, and Antonio M. Anza. The revolution did not put an end to the use of "scientific" prescriptions to treat social problems; if anything it increased this technocratic trend. The torch of technocracy had been carried forward by students of the Porfirian era who transitioned from a more strictly positivist, *científico* approach to progressive concepts of governance, industrialization, and advancement that still promoted individualism and liberal economics but with increased attention to social welfare.

Rolland had run into many roadblocks and failures during his attempt to create his vision of a progressive and modern Mexico, but if the politics of the revolution ultimately left him and many of his associates jaded, they were nonetheless still responsible for creating significant change. It is true that a number of the projects that had consumed Rolland—the free ports and the associated projects of the ship railway and stationary dredge, as well as the single-land tax—were not successful. Populist politics, in his mind, had derailed more "scientific" approaches to economics, municipal governance, and agrarian reform. But the last generation of engineers to come of age during the Porfirian era had had a strong hand in constructing the revolution—they designed and implemented policies on education, infrastructure, resources, industry, and the economy. Politicians and military leaders striving to appease supporters and enemies, to create political stability, and to personally retain power changed and diluted technocratic visions, but it was still largely technocratic visions that made up the foundations of twentieth-century Mexican governance and developmental concepts—a legacy still evident today. Revolutionary leaders wanted Mexico to be connected to the industrialized world, and they wanted to follow through with massive programs that would affect the lives of millions of Mexicans. These goals were (and are) impossible to carry out without trained professionals. As the revolution petered out, many of Mexico's leaders themselves became technocrats; they were often university-trained economists.

For Mexico the shift toward technocratic rule was almost inevitable. Since the initial days of the Industrial Revolution, the tentacles of progress had wrapped more of the world together. Mexico would not and could not escape this change. And while this transformation was (and still is) resisted by many people, most engineers, like most other Mexican intellectuals and political leaders, embraced it as a way to build wealth and Mexican prestige, as well as to protect Mexican sovereignty from more powerful and technologically advanced nations. During the 1940s and 1950s powerful critics wrote the revolution's obituary: a new generation had taken over. But the industrializing, capitalist, and middle-class policies that Alemán, Ruiz Cortines, and López Mateos had put in place were not directly opposed to the revolution; they represented the continuation of particular threads of the revolution that had long existed within bureaucratic halls and in the blueprints of engineers.

Conclusion

Final Thoughts about Modesto Rolland's Life and Legacy

People continue to gather and cheer in Xalapa Stadium and Plaza de Toros, but today few Mexicans have heard of Modesto Rolland—the bureaucrat, Mr. Bothersome, the little-known polymath engineer who cared deeply for Mexico. He had sacrificed much in his attempt to build his vision of a better world in the face of creeping public doubts about equating development with progress. Surely even Rolland during his bouts of sleeplessness had doubted whether his not always successful projects had been truly good for Mexico.

Yet Rolland, like many other middle-class, progressive, liberal Mexicans of his time, genuinely believed for most of his life that Mexicans had to embrace the modern world, and on their own terms as much as possible, if they were to increase prosperity, create greater political and cultural unification, and defend Mexico from unavoidable and powerful outside forces. In Mexico during the first half of the twentieth century that meant dealing with a complicated and violent revolution, cultural heterogeneity, and struggles over resources that played out between developers, *hacendados*, and indigenous communities that still operated to some extent in noncapitalist societies but had been forced into relationships with an insatiable global capitalist order. It meant dealing with an immensely powerful northern neighbor and a frenzy over oil and the rapid expansion of cars, planes, ships, and wars. It meant facing the rise of broadcast media and the resulting expansion of nationalism, international marketing, and global cultural exchange. Many of Rolland's undertakings themselves focused on how to transform projects initially led or influenced by foreign specialists and entrepreneurs—the railroads, ports, and the development of Mexico's peripheries—into enterprises controlled by Mexicans. Adapting ideas and technologies from around the world, especially the West, Rolland strove to create a more secure Mexican sovereignty and to increase Mexican unity through infrastructure, shared space, the adoption of foreign ideas, and the creation of nationalist sentiment.

Rolland did a lot in his storied career; some aspects of his legacy are easier to untangle than others, but few of them are uncomplicated. It is my goal here to pick at those tangles just a little more, one last time. I am also going to make a few concluding statements about the aspects of Rolland's career that stand out as particularly significant to me and useful for revealing larger trends in Mexican and global history.

Rolland's posterity, his living legacies, are themselves fairly diverse. They live in Mexico and the United States, exemplifying Rolland's own back-and-forth relationship between the neighboring countries. He was a Mexican nationalist, but he was a product of Mexico, the United States, and a global movement of people, things, and ideas. In 2011 the descendants of Modesto and Virginia met with the descendants of Modesto and Rosario at a reunion in Querétaro, which I attended after first meeting Rolland's grandson Jorge M. Rolland Constantine through the internet. They shared an appreciation for Modesto and his work. They recognized his less likable attributes but focused more on his achievements. Modesto's descendants are particularly similar in that for the most part they live comfortable, middle-class lives.

As for his material legacy, Rolland was clearly one of the early innovators of reinforced-concrete construction in Mexico. He lectured widely on the subject, registered a number of patents, and built concrete buildings in Mexico City, Cuernavaca, and Mérida. Many of these buildings still stand. He designed multiple stadiums, port works, and water systems. Concrete architecture has received a mixed welcome from designers around the world, but it is difficult to imagine Mexico today without it. It has become far and away the most common building material for housing in the country, especially in places with limited access to wood. Millions of people today live in concrete homes.

Rolland's work during the most militaristic stages of the revolution is less physically visible but no less important. Rolland was an instrumental player in U.S.-Mexican foreign relations during the revolution. He, more than just about any other Mexican, shaped U.S. opinion about the Constitutionalists. This was especially true among progressive intellectuals. Rolland's part in the debates over key issues during the revolution—agrarian reform, municipal governance, subsoil rights, the petroleum industry, U.S. intervention, communications, and education—has not been well recognized. Even if his vision for agrarian reform and a single-land tax did not win out in the end, at one point they were discussed and influential. The 1915 petroleum commission he was a part of laid the groundwork for policies that had huge ramifications for Mexico. And though it is doubtful that Woodrow Wilson would have gone to war over the 1916 Carrizal incident if Rolland had not been in the United States, Rolland was nonetheless a substantial voice. His actions helped defuse a potentially disastrous situation. Rolland's work as a Constitutionalist agent in the United States provides insights into Mexican foreign policy, the global exchange of ideas, and the Mexican Revolution.

Rolland's involvement in the global progressive world was significant in and of itself. He exemplifies Mexico's participation in an international movement during the late nineteenth and early twentieth century that connected a number of far-flung countries, including places as diverse as England, Germany, New Zealand, Argentina, Brazil, the United States, Mexico, and China.[1] The revolution forced Rolland and a number of other Mexican political exiles and diplomats into greater contact with foreign progressives and their ideas. In turn those figures had a profound impact on the revolution. The Progressive movement, alongside nineteenth-century liberalism, was the most influential intellectual trend among the leadership of the Constitutionalists. It fueled the hurried and loose cohesion of these agents of change who pushed for modernization, a multitude of confusing interpretations of "socialism," municipal reform, and a movement sympathetic to the working class but that focused more on creating a strong, modern, and hygienic middle class. The progressive influence further propelled Constitutionalists toward a moderate reformist approach—changing capitalism instead of replacing it. This close association with progressivism separated the victors of the revolution from other revolutionary movements associated more with communism, in Russia and Cuba, for example, and to some extent other Mexican revolutionaries, such as Emiliano Zapata, who focused locally and as much on the past as on the future. Combined with lingering influences from nineteenth-century liberalism and even Spanish colonial policies, this connection to global progressivism explains why Mexican revolutionary technocrats, and intellectuals more broadly, incorporated such a variety of contentious ideas, including Rolland's obsession with the progressive, land-nationalizing (at least in Rolland's early and not exactly accurate interpretation), and otherwise free-trade policies of Henry George. In the heads and hands of the people who built the structures of revolutionary Mexico, this progressive persuasion left a durable imprint that did not always mesh well with Mexican realities.

There are places where Rolland's legacy has seen a recent revival. The stadium he built in Xalapa still stands, is used often, and remains an emblem of the city.[2] In recent years the stadium has been used for international sporting events, for concerts, and, somewhat ironically, for large gatherings of evangelicals and Catholics. In 2015 Jorge M. Rolland, descendants of William K. Boone and Heriberto Jara, and prominent members of Xalapa's government and the Universidad Veracruzana organized celebratory events to commemorate the stadium's ninetieth anniversary.[3] One of the streets entering the stadium has been renamed for Modesto C. Rolland. Rolland's vision of a community-unifying stadium has in some ways come true.

The fact that the stadium is still standing, and in relatively good shape at that, is a testament to Rolland's design and the work of the people who provided the labor to build it. Officials in La Paz, Baja California Sur, Rolland's hometown, are also considering naming a street for Rolland. He has received attention in the works of journalists and scholars on the peninsula. Jorge M. Rolland published a celebratory biography about his grandfather in 2017.[4]

As for the City of Sports, the Plaza de Toros continues to host events and remains the world's largest bullfighting ring. Once on the outskirts of Mexico City, the City of Sports has since been swallowed by the metropolis. In addition to bullfights, it continues to host boxing matches and popular concerts. Its sister stadium's legacy has been important, too. It will, however, become the first of Rolland's stadiums to fall. For decades it hosted both American-style football and soccer events. It stopped holding American-style football games in 1990. It then became home to the Mexican soccer club Cruz Azul, or Blue Cross, in 1996. Mexico's national soccer team has also played there on a number of occasions. In the summer of 2016 the Cosío family, the stadium's owners, announced that the stadium would be demolished sometime in 2018. Commercial developers plan to build a shopping center and hotel on the stadium site. For the moment there are no plans to tear down the Plaza de Toros, which remains an icon of the city.[5]

Of all the projects in his career, Rolland put more time into developing the free ports than any other. The multidecade, on-and-off-again experiment was one of the greatest frustrations of his life. He had convinced a number of people to use the warehouses in the free ports, but despite his best efforts the operation never created the industrial boom that had been his hope. His goals for improving the lives of Tehuantepec residents fell flat and unappreciated except maybe by planners of developmental schemes who came after him. Most of the local people in the region who knew Rolland and his project didn't care for either of them.

With the privilege of hindsight, we can tease out a number of reasons that help explain the failure of the free ports and their associated projects that might also elucidate broader difficulties in modernizing Mexico. The most straightforward reasons for the failure of the ship railway was its unconventionality and a lack of financing. The government was supposedly willing to put up half the cost, but Rolland does not appear to have obtained the other half, which he hoped to get from U.S. financiers. Some of the inability of the government to fully fund this project stemmed from low levels of tax income. As for U.S. investors, Rolland failed to convince them that his radical idea was sound enough, and interest in the isthmus among U.S. officials and business leaders temporarily waned in the postwar years. There was also opposition from American interests associated with U.S. free ports.

As for the stationary dredge, it never came to be for a number of reasons. Rolland underestimated the problems involved in getting it to function. It was more difficult, costly, and labor intensive than Rolland originally imagined. While sorting out these problems, Rolland faced labor organizations that despised him and the project, accusations of corruption, and changes in political administrations. It was a combination of these barriers that halted the project. We will never know if the project Rolland conceived would have worked as planned.

The biggest problem Rolland faced was a lack of consistent support. There were too many drastic policy changes during the revolution. Weetman Pearson's most recent biographer, the historian Paul Garner, writes that Pearson's completion and profitable operation of the transisthmian railway and its connecting ports before the revolution had more to do with the active promotion of his interests by the Porfirio Díaz administration than with Pearson's exceptional entrepreneurial skills, political influence, or financial muscle.[6] The same logic could be applied to Rolland's less successful experience with the same infrastructure. His failure had more to do with inconsistent support and assistance from the revolutionary and postrevolutionary governments than any lack of skill. Administrations supported the project, then pulled support, closed the free ports, then reopened the free ports, and ultimately fired Rolland. Multiple shifting and fighting bureaucracies were involved with the operation of the standard-duty ports, free ports, and railway. Money waxed and waned. Worker unrest was the norm. Under Plutarco Elías Calles and the succeeding presidents of the Maximato era (1928–34), the Tehuantepec ports became almost unusable. There was a world war and then a global cold war. Mexico's northern neighbor was no longer just a colossus, it was a titan; when its prominent residents did not support a project in the Western Hemisphere, there were repercussions.

There are other possible explanations, too. Perhaps there was something to one of the early critiques provided by the engineer and *científico* writer Francisco Bulnes: Tehuantepec was not as strategically situated as Rolland, Salvador Alvarado, and Luis Cabrera thought it was. Geographically it was well situated as a center of interoceanic commerce, but the region was rural, poor, and dizzyingly diverse. Many people spoke different languages and embraced more traditional lifeways. There was little industry and little local support. As for the free ports, most people in the region had no idea what they were. Unlike the rural setting of Rolland's free ports, most of those in Europe and the United States were connected to massive industrial cities, the main drivers of capitalist enterprise.[7] Perhaps the region required even greater industrial development, entrepreneurial interest, and infrastructure before something like the free ports could function well. If Rolland had been provided more consistent support from the beginning, before the rise of the U.S. free ports, maybe the outcome would have been different. Again, we'll never know.

In some ways the free ports were not a complete failure. For better and worse, Rolland's experiments in free-trade initiatives were important precedents for the low-tax and free-trade policies currently in place in Mexico, including the North American Free Trade Agreements (NAFTA) and federal tax incentives for industries established in Tehuantepec. It is impossible to know what would have happened if Lázaro Cárdenas had not reopened the free ports and put them under Rolland's leadership, but the ports had definitely fallen into disrepair in the years between 1925 and 1938, when Rolland was not working on them. Rolland did more than most people to keep these ports and the connected railway functional and used. The free ports continued on until the 1970s, when the government closed them down. But Salina Cruz, Coatzacoalcos, and the Tehuantepec Railway have remained important for international trade. There are ongoing and contested projects similar to Rolland's that are under way in the region, and though they are not direct results of Rolland's work—there were decades of different projects and ideas between 1953 and 2018—current initiatives definitely stand in part on Rolland's shoulders.[8]

Not all of the blame for Rolland's professional failures can be placed on others of course. He was stubborn. He rarely played well with others. In attempting to create massive structural and societal changes, Rolland worked up his designs with only a small group of other planners and politicians, never seriously consulting the people whose lives he wanted to change. He was an effective builder and a passionate writer, but he was not effective at communicating with most everyday Mexicans in the communities where he worked. He didn't talk to them much and he didn't really listen to them. Because of this fault, he often failed to gain the support of locals in the communities where he worked; as a result, many of his projects met with resistance and he grew bitter. This trait was not Rolland's alone; other engineers and technocrats were similar.

One clearly negative legacy of development in Tehuantepec and elsewhere in Mexico has been environmental degradation. The Coatzacoalcos region, especially the river, has become seriously polluted, mainly from the petroleum industry. There have been moves to increase regulatory oversight, but with more pipelines, oil production, sulfur and iron mining, and industrial projects in the works, it seems unlikely that there will be a reversal any time quickly in the trend of environmental deterioration that has persisted for more than a century. The isthmus contains an immense amount of biodiversity threatened by continued development.

Rolland was not ignorant of the fact that humans caused environmental harm, though he had no idea about how serious environmental problems would become by the end of the twentieth century. And although Rolland argued that Mexicans had to exploit their natural resources for their benefit, he was not against sustainable practices. He complained about the misuse of forests and the environmental damage done by certain forms of net fishing in the Sea of Cortez. The latter practice the Mexican government has greatly curtailed while creating protected marine parks and supporting certain commercial fisheries and sport fishing. But Rolland was no environmental activist. He was a fervent promoter of increased petroleum production, railroad construction, and industrialization. To him, as with many business leaders and government officials today, there was a balance to be struck. But more often than not that balance has not been maintained, a reality that will have to change moving forward if Mexico wants to develop a sustainable environment and a better life for its people.

Much of Rolland's legacy is still unfolding. In some ways it is wrapped up in people and trends that stem from his career but also transcend it. His life is part of the larger processes of modernization and globalization. If humans develop a more interconnected and technologically advanced global society that benefits the majority of the world's people without destroying the planet, then people who look back at Rolland's life may very well see him in a light that would have been much to his own liking—that he helped bring about social and material betterment. His continuing legacy will also depend on whether those benefits—less want, longer lives, improved access to quality education, increasingly useful technologies—outweigh the new problems these developments produce. For Rolland the ardent nationalist, his legacy depends too on Mexico's ability to meet these high standards. Extreme poverty around the world, including Mexico, has declined in recent decades.[9] However, inequality among the richest and poorest people on earth and among the richest and poorest nations on earth has risen to alarmingly high levels.[10] Mexico remains an example of inequality at its worst. All too often local residents are still left out of decision-making. If inequality continues to cause severe disorder, or if environmental catastrophes stemming from global warming and pollution reach critical levels and drastically harm the human race, then many people will likely see Rolland's legacy as one of well-intentioned but disastrous short-sightedness.[11] Whether that view is fair or not, he will be seen as one of many drivers of Faustian development who destroyed the world in the name of improvement.

It was somewhere halfway through writing this book that I realized just how ambivalent I had become about Rolland. Like Rolland, I want to believe in humanity's ability to cooperate and continuously build a better world, but I have my doubts about the sustainability of the current global capitalist order that Rolland criticized but helped create. I am concerned about many of progress's more unsightly children: greed, inequality, hierarchical power structures, pollution, and an obsession with material wealth, economic growth, and comfort. Looking back after completing this book, I know now that it was my own ongoing ambivalence about progress—my experience of modernity—in addition to Rolland's complexity, that influenced me to write in the introduction to this book that Rolland was an ambiguous figure. That has not changed. I think societies benefit from optimistic idealists, scientific and otherwise. But the lives of Rolland and others like him teach us that a certain level of cynicism and doubt are also important. All dreams are potential nightmares, and large projects sold as beneficial almost always come with real costs.

The creation of modern Mexico, the Mexico that Rolland helped forge out of the smoke and fire of the revolution, has been a painful process for many Mexicans. Rolland correctly understood that there was no turning back to premodernity and no way to avoid the world rumbling around and through Mexico. He strove to use modern forces to improve his country. It is what drove him. But he often took it upon himself to know what was best for those who failed to grasp the impending disruptions of progress. He attempted to force his will on people who understood his designs but had no desire to participate. Rolland's bullheadedness upset his political superiors and employees alike. He sometimes underestimated the forces he came up against, as well as the difficulty of creating a happier, more prosperous Mexico. He failed to see the extent of the environmental ramifications and the emotional ambivalence often caused by modern living.

Yet Rolland also helped build a Mexico with increased wealth, interconnectedness, and a larger middle class. Even though he did not care for the outcome of revolutionary agrarian reform, he and the associates with whom he tangled helped initiate policies that reshaped peoples' connection to the earth, each other, and their government. He constructed stadiums that brought people together in celebration and that became globally recognized symbols of art, sport, and architecture. Rolland was a conduit for the international exchange of ideas and a promoter of transnational education. He and his associates built an infrastructure that has expanded access to health services, schools, and material goods. Rolland was a key figure in unifying Mexico into a more coherent nation. He connected it through infrastructure and as a prominent intermediary in Mexico's interaction with larger global forces. The strong-willed son of a French adventurer and a Baja California woman who believed in the power of education, Rolland became the face of change in Mexico—a harbinger of modernity, an apostle of progress.

Notes

Introduction

1. This view was heavily influenced by Steinbeck's longtime friend and co-creator of *The Log from the Sea of Cortez*, the biologist Ed Ricketts.

2. After embarking on his voyage around the Baja California peninsula, which provided the material for *The Log from the Sea of Cortez*, Steinbeck moved on to help produce *The Forgotten Village*, a documentary that supported government-financed education and modern medicine. It also condemned the ignorance of rural natural healers. Ricketts detested this turn in Steinbeck's thought. This story is covered in Warfield, "Steinbeck and the Tragedy of Progress," 102–6.

3. Bijker, Hughes, and Pinch, *Social Construction of Technological Systems*, 9.

4. Josephson, *Industrialized Nature*, 8.

5. Rodgers, *Atlantic Crossings*, 277–87.

6. Examples of exceptional historical works on Mexico that have influenced me and that are either biographies or use specific people as a narrative entry into broader themes include the works of Vanderwood, especially *Power of God against the Guns of Government*; Womack, *Zapata and the Mexican Revolution*; Garner, *British Lions and Mexican Eagles*; Fowler, *Santa Anna of Mexico*; Paxman, *Jenkins of Mexico*; and Jacoby, *Strange Career of William Ellis*.

7. Kloppenberg, *Uncertain Victory*, 299.

8. Some of the most prominent works that address this global progressive movement are Rodgers, *Atlantic Crossings*; Kloppenberg, *Uncertain Victory*; Tyrrell, *True Gardens of the Gods*; Coleman, *Progressivism and the World of Reform*; and Fischer, *Fairness and Freedom*.

9. Some of these works include Britton, *Revolution and Ideology*; Delpar, *Enormous Vogue of Things Mexican*; Hale, "Frank Tannenbaum and the Mexican Revolution"; Hale, *Emilio Rabasa*; Brading, *Prophecy and Myth in Mexican History*; Brading, *Caudillo and Peasant*; Hamon and Niblo, *Precursores de la revolución*; Ruiz, *Mexico*; R. Flores, *Backroads Pragmatists*; Tenorio-Trillo, "Stereophonic Scientific Modernisms"; and Tenorio-Trillo, *I Speak of the City*.

10. Some works that have addressed the issue of revolutionary factions having agents in the United States include Richmond, *Venustiano Carranza's Nationalist Struggle*; Raat, *Revoltosos*; Harris and Sadler, "'Underside' of the Mexican Revolution"; M. Smith, "Carrancista Propaganda and the Print Media"; M. Smith,

"Mexican Secret Service in the United States"; Anderson, *Pancho Villa's Revolution by Headlines*; and Britton, *Revolution and Ideology*.

11. Works that touch on engineers, development, and technology in modern Mexico include Coatsworth, *Growth against Development*; Tinajero and Freeman, *Technology and Culture*; Agostoni, *Monuments of Progress*; A. Alexander, *City on Fire*; Connolly, *El contratista de don Porfirio*; Saldaña, *Las revoluciones políticas y la ciencia*; Wolfe, *Watering the Revolution*; Wakild, "Naturalizing Modernity"; Beatty, *Technology and the Search for Progress*; and Hill, "Circuits of State."

U.S.-published English-language histories specifically on Mexican engineering have not existed to date. For Mexican authors' Spanish-language works, see Domínguez Martínez, *La ingeniería civil en México, 1900–1940*; de Ibarrola, *Apuntes*; Tamayo, *Breve reseña sobre la Escuela Nacional de Ingeniería*; Aboites Aguilar, *Pablo Bistráin*; de la Paz Ramos Lara and Rodríguez Benítez, *Formación de ingenieros*; Bazant, "La enseñanza y la práctica de la ingeniería durante el Porfiriato"; and de Gortari Rabiela, "Educación y conciencia nacional."

12. Histories of the Mexican Revolution abound. Some popular works include Matute, *Historia de la Revolución Mexicana, 1917–1924*; Knight, *Mexican Revolution*; Ruíz, *Great Rebellion*; J. Hart, *Revolutionary Mexico*; Katz, *Life and Times of Pancho Villa*; Marván Laborde, *La Revolución Mexicana, 1908–1932*; Gonzales, *Mexican Revolution*; and Joseph and Buchenau, *Mexico's Once and Future Revolution*.

13. Historians who have delved into this murky world of the middle and the lives of Mexican technocrats in works I build on include Cockcroft, *Intellectual Precursors*; Shadle, *Andrés Molina Enríquez*; Craib, *Cartographic Mexico*; Babb, *Managing Mexico*; L. Carranza, *Architecture as Revolution*; Gauss, *Made in Mexico*; Ervin, "Formation of the Revolutionary Middle Class"; and Eineigel, "Revolutionary Promises Encounter Urban Realities."

1. Child of the Porfiriato, of the Periphery

1. "The Largest Pearl Ever Found," *Washington (DC) Bee*, April 14, 1883, 4.

2. "Pacific Coast Pearls," *Washington Bee*, January 10, 1885, 1; Cariño and Monteforte, "History of Pearling," 95.

3. Cariño and Monteforte, "History of Pearling," 95.

4. "Juan Francisco Rolland y Ma. Jesús Mejía," ACNSLP, Archivo Eclesiástico; Domínguez Tapa, *Forjadores*, 207.

5. "Compulsory Education, Says Carranza, Is the Cure for Mexico's Ills," *The Sun* (New York), September 20, 1914, sec. 3, 8.

6. According to a report by the Secretaría de Fomento, or Ministry of Development, in December 1887 La Paz had 6,463 residents. "Secretaría de Fomento," *Diario Oficial*, December 9, 1887, in NARA, RG 59, vol. 94, *Despatches from United States Ministers to Mexico, 1823–1906*, microfilm roll 89, September 1–December 31, 1887.

7. Rivas Hernández, "La industría," 292, 308–9, 323; "The Mines of Mulege," *Los Angeles Times*, May 31, 1883, 2.

8. Adas, *Machines as the Measure of Men*.

9. The idea that telegraphy, and electricity more generally, works as an extension of the individual nervous system is in McLuhan, *Understanding Media*, 248. Also see Noyola, *La raza de la herba*.

10. Buffington and French, "Culture of Modernity," 410. For more on education during the Porfirian era, see Vaughn, *State, Education, and Social Class in Mexico*.

11. Ibarra Rivera, *Historia de la educación en Baja California Sur*, 106, 140, 144, 158; Domínguez Tapa, *Forjadores*, 207–8; Amelia Rolland to Jesús García, January 31, 1885, AHPLM, IP, doc. 32, vol. 191, exp. s/n; Clemente Trujillo to Jefe Político y Comandante Militar del Territorio, August 24, 1886, AHPLM, IP, doc. 218, vol. 200, leg. 8, exp. s/n.

12. See M. Bernstein, *Mexican Mining Industry*.

13. Ferguson, *House of Rothschild*, 481–85. Records of the ships that entered the Santa Rosalía port are in the AHADB.

14. Preciado Llamas, *En la periferia del régimen*, 24–25; Garner, *Porfirio Díaz*, 68, 90. The term "tentacles of progress" comes from Headrick, *Tentacles of Progress*.

15. "Compulsory Education, Says Carranza, Is the Cure for Mexico's Ills," *The Sun*, September 20, 1914, sec. 3, 8.

16. "Compulsory Education, Says Carranza, Is the Cure for Mexico's Ills," *The Sun*, September 20, 1914, sec. 3, 8.

17. Certificate of completion of a teacher's degree, December 16, 1903, JMR; Jorge M. Rolland C., email to J. Justin Castro, January 20, 2015; Rodríguez Benítez, "La formación de ingenieros en el Colegio Rosales," 131–35.

18. J. Rolland, "Estudios de Modesto C. Rolland," 1–4.

19. For more on how railroads changed perceptions of space and time, see Schivelbusch, *Railway Journey*. For Mexico specifically, see Matthews, *Civilizing Machine*, 55–101.

20. "Fashionable," *Mexican Herald*, January 24, 1901, 2.

21. For an examination of Porfirian consumer culture, see Bunker, *Creating Mexican Consumer Culture*.

22. J. Rolland, "Estudios de Modesto C. Rolland," 1–4; J. Rolland, *Modesto C. Rolland*, 21.

23. Secretaría, Instituto Científico y Literario del Estado de Hidalgo, "Modesto C. Rolland y Julián Adano: El gobierno les concede examen de nociones de cálculo infinitesimal," AGEH, AH, fondo Instituto Científico y Literario del Estado, exp. alumnos, Modesto C. Rolland.

24. J. Rolland, "Estudios de Modesto C. Rolland," 1–4. Rolland's educational records are housed at the Archivo Histórico de la Universidad Nacional Autónoma de México in Mexico City.

25. Tenorio-Trillo, *I Speak of the City*, 284–85. For a discussion about the rise of technocrats, including engineers, in the late Porfirian era, see Gauss, *Made in Mexico*, 1–23.

26. Candiani, *Dreaming of Dry Land*, 153–55.

27. For a study of Mexico's dependence on foreign expertise and technology in the nineteenth century, see Beatty, *Technology and the Search for Progress in Modern Mexico*.

28. Bazant, "La enseñanza y la práctica de la ingeniería durante el Porfiriato," 255–86; A. Alexander, *City on Fire*, 58; Domínguez Martínez, *La ingeniería civil en México*, 31.

29. Domínguez Martínez, *La ingeniería civil en México*, 45–46.

30. Domínguez Martínez, *La ingeniería civil en México*, 45–46.

31. "Notes," *Annals of the American Academy of Political and Social Science* 1 (April 1891): 705–7.

32. Pani, *Mi contribución*, 138–39.

33. A. Alexander, *City on Fire*, 59–60.

34. Records on Rolland's work at the National Agriculture and Veterinary School are located in the AGN, IP, caja 216, exp. 17, f. 19, and caja 17, exp. 52, f. 10. See also "Por la Escuela de Agricultura," *El Agricultor Moderno*, January 1, 1905, 10.

35. Rolland, *Lecciones sobre presas*; Rolland, *Algunas lecciones sobre el levantamiento de polígonos por "deflexiones."*

36. Elena de la Garza de Meléndez, "Jorge Rolland de la Garza," copy provided to the author by Deanna Wicks, a granddaughter of Modesto and Virginia. Elena de la Garza was the sister of Virginia.

37. Sierra's speech quoted in de la Garza de Meléndez, "Jorge Rolland de la Garza."

38. De la Garza de Meléndez, "Jorge Rolland de la Garza"; Deanna Rolland Wicks, email to Harry W. Schroeder Jr. and Justin Castro, April 2, 2015. Wicks, as noted earlier, is the granddaughter of Modesto and Virginia. Schroeder is the grandson of Elena de la Garza.

39. Jorge M. Rolland, email to J. Justin Castro, March 11, 2015.

40. Deanna Rolland Wicks, email to Harry W. Schroeder Jr. and Justin Castro, April 2, 2015; Jorge M. Rolland, email to Justin Castro, April 4, 2014; J. Rolland, "Notas de la vida de la familia Rolland Garza," 2014, unpublished document, JMR.

41. Virginia contra Modesto Rolland, divorcio, April–June 1909, AGN, TSJDF, folio 148123, 17 fojas.

42. Modesto Rolland to Virginia Garza de Rolland, Virginia contra Modesto Rolland, divorcio, April–June 1909, AGN, TSJDF, folio 148123, 17 fojas; Deanna Wicks, email to J. Justin Castro and Jorge M. Rolland, April 26, 2015.

43. A. Alexander, "Safety by Design," 444. Also see Agostoni, *Monuments of Progress*, 115–16; and Wakild, "Naturalizing Modernity," 101–4.

44. Scott, *Art of Not Being Governed*, 8, 13.

45. Vice president of Cia. de Cemento Portland "La Tolteca" to L. Salazar, December 16, 1910, AHPM.

46. These patents are located in the AGN, P.

47. Tafunell, "On the Origins of ISI," 302–8.

48. A. Alexander, *City on Fire*, 57–73.

49. Wakild, "Naturalizing Modernity," 119. For more on the Grand Canal and drainage system of Mexico City, see Garner, *British Lions and Mexican Eagles*, 62–93; and Agostoni, *Monuments of Progress*, 61–93.

50. Modesto C. Rolland, "Memoria descriptiva de las obras de provisión de aguas potables para la Ciudad de México," 1909, AHPM, exp. 5, Modesto C. Rolland; Marroquín y Rivera, *Memoria*. This *memoria*, along with many letters from Marroquín y Rivera, Rolland, and other engineers and engineering students who worked on this project, are located at the AHPM.

51. Marroquín y Rivera, *Memoria*, 585–87.

52. Cámara Nacional del Cemento, *Medio siglo de cemento en México*, 7; J. Rolland, "Ingeniero Modesto C. Rolland," 41; "La introducción de aguas potables a la capital," *El País* (Mexico City), September 23, 1910, 1, 7. For a solid piece on the history of the Xochimilco waterworks, see Banister and Widdifield, "Debut of 'Modern Water' in Early 20th Century Mexico City."

53. M. Rolland, *Cemento armado*.

54. In particular, students and professors in law, journalism, and art associated with the group Ateneo de la Juventud became opposed to positivist doctrines and discussed works by philosophers, literary figures, and activists who attacked the Enlightenment's extreme rationalism. See Garciadiego Dantan, *Rudos contra científicos*, 49–65.

55. Garner, *Porfirio Díaz*, 185.

56. M. Rolland, "Investigation Work into the Municipal City Governments," 107. For more on Rolland's work in the Engineers' Club, see M. Rolland, "Agrarian Problem in Mexico."

57. J. Hart, *Empire and Revolution*, 122.

58. "Railroad Development in Mexico," *Dun's Review*, January 1905, 23–34; "Brown Returns to Mexico," *New York Times*, September 13, 1913, 3.

59. J. Hart, *Empire and Revolution*, 106–30.

60. "Railroad Development in Mexico," 23–24; Alzanti, *Historia de la mexicanización de los Ferrocarriles Nacionales de México*, 162–64.

61. "Entran a la lucha nuevas energía," *El Dictamen* (Veracruz), May 25, 1909, 1.

62. M. Rolland, "Investigation Work into the Municipal City Governments," 106.

63. M. Rolland, "Investigation Work into the Municipal City Governments," 106.

2. The Reluctant Revolutionary

1. Díaz's actual birthday was September 14. "Mexican Historic Pageant," *New York Times*, September 16, 1910, 6; "Ambassadors Received by President Diaz," *Atlanta Constitution*, September 6, 1910, 12; Tenorio-Trillo, "1910 Mexico City," 77.

2. Joseph and Buchenau, *Mexico's Once and Future Revolution*, 33–35; Ruíz, *Great Rebellion*, 120–35.

3. Knight, *Mexican Revolution*, 1:75.

4. Jorge M. Rolland, email to Justin Castro, May 6, 2015; J. Rolland, "Ing. Modesto C. Rolland Mejia: Antecedents," 2013, unpublished document, JMR.

5. For more on the Monterrey business elite and the Mexican state, see Saragoza, *Monterrey Elite*.

6. Gilly, *Mexican Revolution*, 54; Garner, *Porfirio Díaz*, 109–10; Knight, *Mexican Revolution*, 1:37–77.

7. Knight, *Mexican Revolution*, 1:62.

8. Knight, *Mexican Revolution*, 1:59.

9. Joseph and Buchenau, *Mexico's Once and Future Revolution*, 34–35; Cumberland, *Mexican Revolution*, 116.

10. Knight, *Mexican Revolution*, 1:77; Gilly, *Mexican Revolution*, 55.

11. Knight, *Mexican Revolution*, 1:175; also see Henderson, *Worm in the Wheat*, 37–41.

12. Knight, *Mexican Revolution*, 1:175.

13. Knight, *Mexican Revolution*, 1:172–88; Joseph and Buchenau, *Mexico's Once and Future Revolution*, 43–45; Gonzales, *Mexican Revolution*, 78–80; Stephan Bonsal, "Mexican Catholics Plan to Rule Nation," *New York Times*, May 23, 1911, 1; "El Sr. Gral. Díaz recibe noticia del temblor," *El Imparcial* (Mexico City), June 10, 1911, 1.

14. M. C. Rolland to Manuel Urquidi, January 13, 1912, NLBL, UP, box 1, f. 3a–3c; M. C. Rolland, "Patente 13832: Un tinaco o tanque de cemento armado," January 9, 1913, AGN, P, exp. 84, leg. 292.

15. "Los ingenieros han formado un club político," *Nueva Era* (Mexico City), August 16, 1911, 3; "Club Antireeleccionista 'Francisco Díaz Covarrubias," *Nueva Era*, August 22, 1911, 5.

16. Many of these same engineers would continue to promote campaigns to put Mexican engineers in charge of large infrastructural projects in the Centro de Ingenieros, which Rolland and others formed in 1918. Specifically they spoke out against hiring foreign firms to build infrastructural and irrigation projects during the 1920s. See Baptista González and Saldaña, "La participación política y reivindicación gremial del Centro de Ingenieros de México," 1221–30.

17. "La mexicanización en las Líneas Nacionales," *Elektron* (Mexico City), September 1, 1911, 3.

18. "Club Progresista Californiano," *Nueva Era*, August 16, 1911, 4.

19. Henry Lane Wilson to Philander C. Knox, "Memorandum Relative to the Mexicanization of the National Railways of Mexico," January 16, 1912, in U.S. State Department, *Papers Relating to the Foreign Relations of the United States*, 914–15; Alzanti, *Historia de la mexicanización*, 41. For more on Mexican railway workers see Alegre, *Railroad Radicals*.

20. "Los ingenieros han formado un club político," *Nueva Era*, August 16, 1911, 3; "La Mexicanización de las Líneas Nacionales," *El Imparcial*, September 22, 1911, 5; "Mexicanización de las Líneas Nacionales," *Diario del Hogar* (Mexico City), September 22, 1911, 1; "La Mejicanización de la Ferrocarriles Nacionales," *El País* (Mexico City), October 16, 2011, 5.

21. "La Mexicanización de los ferrocarriles," *El Tiempo* (Mexico City), October 4, 1911, 6; "La Mexicanización de los ferrocarriles," *El Imparcial*, October 4, 1911, 7.

22. "Mexicanización de los Ferrocarriles Nacionales," *Nueva Era*, October 30, 1911, 5.

23. "La Mexicanización de los ferrocarriles," *Diario del Hogar*, October 26, 1911, 1.

24. "Meetings: National Railways of Mexico," *Wall Street Journal*, October 6, 1911, 7.

25. "Pres. Brown on Prospects on Railways of Mexico Territory," *Wall Street Journal*, October 18, 1911, 4.

26. "Normal Conditions in Mexico," *Wall Street Journal*, July 13, 1911, 8 (quoted text); "National Railways of Mexico," *Wall Street Journal*, November 9, 1911, 5.

27. "The Railway Situation in Mexico," *Railway World* 55, no. 2 (August 11, 1911): 662.

28. M. Rolland, "Investigation Work into the Municipal City Governments," 107.

29. "La Mexicanización de las Líneas Nacionales," *Diario del Hogar*, November 3, 1911, 4. For more on the larger Mexicanization movement, see Alzanti, *Historia de la mexicanización*.

30. Henry Lane Wilson to Philander C. Knox, January 16, 1912, in U.S. State Department, *Papers Relating to the Foreign Relations of the United States*, 914.

31. Jaime Gurza, *The Railway Policy in Mexico*, quoted in U.S. State Department, *Papers Relating to the Foreign Relations of the United States*, 1912, 912.

32. "National Rys. of Mexico in Danger of a Strike," *Wall Street Journal*, April 11, 1912, 5; "American Engineers Out on National of Mexico System," *Wall Street Journal*, April 19, 1912, 7; "Process of Mexicanization," *Wall Street Journal*, May 13, 1912, 2.

33. Henry Lane Wilson to Philander C. Knox, April 2, 1912, in U.S. State Department, *Papers Relating to the Foreign Relations of the United States*, 918.

34. "American Engineers Out on National of Mexico System," *Wall Street Journal*, April 19, 1912, 7.

35. Philander C. Knox to the Presidents of the eighty-six companies of the United States operating lines of five hundred miles or more, May 17, 1912, in U.S. State Department, *Papers Relating to the Foreign Relations of the United States*, 923; William H. Taft to [F. M.] Huntington Wilson, May 12, 1917, in U.S. State Department, *Papers Relating to the Foreign Relations of the United States*, 922–23; "Taft to Engineers' Aid," *New York Times*, May 20, 1917, 1.

36. For histories of the Morelos revolutionaries, see Womack, *Zapata and the Mexican Revolution*; Samuel Brunk, *¡Emiliano Zapata!*; and P. Hart, *Bitter Harvest*.

37. The Zapata and Orozco rebellions are covered in a number of books about the Mexican Revolution. I based much of this and the preceding paragraph on Joseph and Buchenau, *Mexico's Once and Future Revolution*, 49–53.

38. Ross, *Francisco I. Madero*, 211–72; Cumberland, *Mexican Revolution*, 200–243.

39. M. Rolland, "Investigation Work into the Municipal City Governments," 106–7.

40. T. K. Eccles and J. F. Ealy to Henry Lane Wilson, April 16, 1912, in U.S. State Department, *Papers Relating to the Foreign Relations of the United States*, 922.

41. "Manifesto a la nación de la 'A.D.P.N.,'" *La Patria* (Mexico City), May 4, 1912, 2.

42. "Llamamiento patriótico a todos los mexicanos," *Nueva Era*, April 28, 1912, 4; "Llamamiento patriótico a todos los mexicanos," *Diario del Hogar*, April 28, 1912, 4.

43. Ross, *Francisco I. Madero*, 276–311; Cumberland, *Mexican Revolution*, 229–43.

44. Pani, *Mi contribución*, 153–75.

45. Carlo di Fornaro, "The Great Mexican Revolution," *Forum*, November 1915, 534.

46. M. Rolland, "Investigation Work into the Municipal City Governments," 108.

3. A Mexican Progressive

1. M. Rolland, "Investigation Work into the Municipal City Governments," 108.

2. M. Rolland, "Investigation Work into the Municipal City Governments," 108.

3. M. Rolland, "Investigation Work into the Municipal City Governments," 109.

4. Martín Luis Guzmán's historical novel *El águila y la serpiente* (1928) provides a well-written, partly fictionalized account of the specialists who fled to the United States and then subsequently united along the Mexican-U.S. border to fight Huerta. The book was translated into English as *The Eagle and the Serpent*.

5. M. Meyer, *Huerta*, 68.

6. Katz, *Life and Times of Pancho Villa*, 200–203. Works that focus on Carranza's government include Taracena, *Venustiano Carranza*; Richmond, *Venustiano Carranza's Nationalist Struggle*; and Krauze, *Venustiano Carranza*.

7. Rolland appears to have first picked up English while at the Colegio Rosales, where he excelled in English courses.

8. M. Rolland, "Investigation Work into the Municipal City Governments," 109–10; "Compulsory Education, Says Carranza, Is the Cure for Mexico's Ills," *The Sun* (New York), September 20, 1914, sec. 3, 8.

9. M. Rolland, "Investigation Work into the Municipal City Governments," 109–10.

10. M. Smith, "Carrancista Propaganda and the Print Media in the United States," 155–74.

11. M. Rolland, "Investigation Work into the Municipal City Governments," 110; "Compulsory Education, Says Carranza, Is the Cure for Mexico," *Baltimore Sun*, September 27, 1914, 7.

12. M. Rolland, "Investigation Work into the Municipal City Government," 110; "Education Notes," *New York Times*, May 28, 1914, 21.

13. "War with Mexico Averted," *Appendix to the Congressional Record*, 63rd Cong., 2nd sess. (Washington DC: Government Printing Office, 1914), 1133; "Many Reforms Needed," *Arizona Sentinel* (Yuma), August 20, 1914, 4.

14. "Wilson's Stand Correct," *Bemidji (MN) Daily Pioneer*, August 17, 1914, 1; "The Reason Why," *The Menace* (Aurora MO), August 22, 1914, 3; "Compulsory Education, Says Carranza, Is the Cure for Mexico's Ills," *The Sun*, September 20, 1914, sec. 3, 8.

15. M. Rolland, "Investigation Work into the Municipal City Governments," 111–13.

16. Coleman, *Progressivism and the World of Reform*, 5; Fischer, *Fairness and Freedom*, 296–97.

17. Fischer, *Fairness and Freedom*, 296.

18. Haley, *Revolution and Intervention*, 114–22.

19. Knight, *Mexican Revolution*, 1:152–55; Enrique Krauze, "The April Invasion of Veracruz," *New York Times*, April 20, 2014, http://www.nytimes.com/2014/04/21/opinion/krauze-the-april-invasion-of-veracruz.html?_r=0.

20. "Entire Surrender Is Only Condition," *Daily Missoulian*, July 20, 1914, 1; "U.S. Still Waiting Carranza's Reply," *New York Tribune*, July 19, 1914, 3.

21. "Urquidi Admits Friction," *New York Times*, June 17, 1914, 1, 3.

22. Isidro Fabela to Modesto Rolland, June 26, 1914, ASRE, exp. 5-17-19.

23. Mexican Bureau of Information, *"Red Papers,"* 1; "Rolland Assails Moheno," *New York Times*, July 12, 1914, 2.

24. "No Truce with Carbajal," *New York Times*, July 17, 1914, 2.

25. Rolland, quoted in Yankelevich, "En la retaguardia de la Revolución Mexicana," 41.

26. Circular, unknown author, September 9, 1914, ASRE, exp. 17-20-33; [name illegible] to Francisco Urquidi, September 18, 1914, ASRE, exp. 17-20-33.

27. For a representative sample see "Entire Surrender Is Only Condition," *Daily Missoulian*, July 20, 1914, 1; "Text of Carranza's Letter Made Public," *Ogden (UT) Standard*, July 20, 1914, 8; "Wilson's Stand Correct," *Bemidji Daily Pioneer*, August 17, 1914, 1; "Mexican Policy an Asset," *Rock Island (IL) Argus*, August 17, 1914, 4.

28. Modesto C. Rolland, "Who Are the Great Generals of the Constitutionalist Army of Mexico?," *Arizona Sentinel*, August 20, 1914, 1.

29. Mexican Bureau of Information, *"Red Papers,"* 1–16.

30. Ignacio Bonillas, Aviso del nombramiento de oficial mayor de la Sria. de Comunicaciones a Modesto C. Rolland, November 7, 1914, AGN, IP, caja 326, exp. 3.

31. "Notes of the Passing Day," *Mexican Herald* (Mexico City), October 12, 1914, 6; "National Railways of Mexico," *Wall Street Journal*, October 15, 1914, 5.

32. "National Railways of Mexico," *Wall Street Journal*, October 29, 1914, 7.

33. Shadle, *Andrés Molina Enríquez*, 1–2, 9, 22–26.

34. Kroeber, *Man, Land, and Water*, 33–56, 186–88.

35. Coleman, *Progressivism and the World of Reform*, xi. See also Fischer, *Fairness and Freedom*.

36. M. Rolland, *Distribución de las tierras*, 1.

37. M. Rolland, *Agrarian Question*, 5–7. For specifics on New Zealand reform policies in the 1890s, see Coleman, *Progressivism and the World of Reform*; Fischer, *Fairness and Freedom*; P. Mein-Smith, *Concise History of New Zealand*; and Brooking, *History of New Zealand*.

38. M. Rolland, *Distribución de las tierras*, 12–13.

39. Coleman, *Progressivism and the World of Reform*, 24, 88–89. In the 1990s officials in Estonia revived an attempt to enact Georgist policies. For more on the influence of George, see Laurent, *Henry George's Legacy in Economic Thought*.

40. "Interesante conferencia en la pepitoria," *Diario del Hogar* (Mexico City), October 23, 1914, 3.

41. "La conferencia de hoy en el Principal," *El Pueblo* (Veracruz), December 23, 1914, 1; "La cuestión agraria ha sido siempre el principal problema de la Revolución," *El Pueblo*, December 24, 1914, 1, 2; M. Rolland, *Agrarian Question*, 9–10.

42. This influential perspective was introduced to U.S. historians in Tannenbaum, *Mexican Agrarian Revolution*.

43. M. Rolland, *Agrarian Question*, 3, 8.

44. Quirk, "Liberales y radicales," 515–16.

45. "Se establecerá un centro de reunión con los elementos revolucionarios," *El Pueblo*, December 25, 1914, 1; "Confederación Revolucionaría dirige una excitativa al pueblo mexicano," *El Pueblo*, January 7, 1915, 1; U.S. Congress, *Investigation of Mexican Affairs*, 2820.

46. Venustiano Carranza, "El C. Primer Jefe de la Revolución expide el primer decreto sobre materia agraria," *El Pueblo*, January 7, 1915, 1.

47. Tannenbaum, *Peace by Revolution*, 162; Shadle, *Andrés Molina Enríquez*, 67–68.

48. Jorge Useta, "Al margen de los sucesos diarios," *El Pueblo*, January 23, 1915, 3.

49. "Se establecerá un centro de reunión con los elementos revolucionarios," *El Pueblo*, December 25, 1914, 1; "El banquete de la aduana," *El Pueblo*, December 30, 1914, 1; "Rumbo al puerto de Tampico salió la Comisión Técnica Petrolífera," *El Pueblo*, January 26, 1915, 1.

50. M. C. Rolland to Jesús Urutea, January 23, 1915, ASRE, exp. 5-7-19.

51. "Rumbo al puerto de Tampico salió la Comisión Técnica Petrolífera," *El Pueblo*, January 26, 1925, 1; "Quedo nombrada la Comisión Técnica del Petróleo, dependiente de la Secretaria de Fomento," *El Pueblo*, March 24, 1915, 1; "Una carta al Sr. Ing. M.C. Rolland," *El Pueblo*, April 14, 1915, 3; "Saldrá para los E.U. la Comisión Técnica del Petróleo," *El Pueblo*, April 26, 1915, 1; "Se embarcó rumbo a Nueva York la Comisión Técnica del Petróleo presidida por el Ing. Rouaix," *El Pueblo*, May 1, 1915, 1; Osorio Marbán, *Carranza*, 85.

52. "Ne se han invadido ningunas atribuciones por las autoridades constitucionalistas," *El Pueblo*, December 9, 1914; "Siguen los trabajos de construcción del muelle fiscal," *El Pueblo*, April 28, 1915, 6; "La meritoria obra hecha en el puerto," *El Pueblo*, June 26, 1915, 3.

53. M. Rolland, "Trial of Socialism in Mexico," 79–90.

4. Back to the Periphery

1. "Se embarcó rumbo a Nueva York la Comisión Técnica Petrolero presidida por el Ing. Rouaix," *El Pueblo* (Veracruz), May 1, 1915, 1; Castañeda Crisolis and Saldaña, "El Boletín del Petróleo"; Michael Alderson, "The Ward Line," accessed October 4, 2015, http://web.archive.org/web/20120716194637/http://www.wardline.com/page/page/4557563.htm.

2. "Johnson and Willard Watched by Crowds," *Atlanta Constitution*, March 29, 1915, 6.

3. "32,000 a Fight," *New York Times*, April 7, 1915, 11.

4. Pérez, *Cuba under the Platt Amendment*, 123–30.

5. "Noticias," *Diario de la Mañana* (Havana, Cuba), May 5, 1915, 8.

6. "Monthly Petroleum Exports Continue on Large Scale," *Wall Street Journal*, May 22, 1915, 6.

7. "United States Must Act at Once on *Lusitania*, Says Colonel Roosevelt," *New York Times*, May 10, 1915, 1.

8. Vázquez Schiaffino, "Memoria relativa al viaje efectuado a los Estados Unidos," 506–34.

9. Saldaña, *Las revoluciones políticas y la ciencia*, 2:252; "Llegan comisionados a Pittsburg [sic]," *Regidor* (San Antonio TX), June 23, 1915, 2.

10. "Studying Oklahoma Oil Fields," *Wall Street Journal*, June 24, 1915, 7; "Japanese Naval Office after Oil Data," *Evening Star* (Washington DC), June 21, 1915, 2.

11. "Oil Developments on the Gulf Coast of Mexico," *Wall Street Journal*, August 20, 1915, 5.

12. "Mexico to Alter Oil Laws," *New York Times*, January 26, 1916, 6. For more on the influence of the Petroleum Technical Commission and Carranza's taxes on petroleum companies, see Richmond, *Venustiano Carranza's Nationalist Struggle*, 97–99.

13. Wasserman, *Pesos and Politics*, 18.

14. Saldaña, *Las revoluciones políticas y la ciencia*, 2:250–53; de Gortari Rabiela, "Educación y conciencia nacional," 125–26.

15. Heriberto Jara, quoted in Zapata Vela, *Conversaciones con Heriberto Jara*, 58.

16. For more on this topic, see Evans, *Bound in Twine*.

17. Joseph, *Revolution from Without*, 93–149.

18. Chacón, "Salvador Alvarado and Agrarian Reform," 189.

19. M. Rolland, "Trial of Socialism in Mexico," 85. On the commission and Rolland's leadership, see "Comisión encabezada por el ingeniero Modesto C. Rolland," *La Voz del Pueblo* (Santa Fe NM), February 3, 1917, 3.

20. Eiss, *In the Name of El Pueblo*, 114.

21. Lewis, "Nation, Education, and the 'Indian Problem,'" 178–80.

22. Dawson, *Indian and Nation*, 6.

23. Tobin, "Insurgent as Ideologue," 327.

24. Record quoted in Tobin, "Insurgent as Ideologue," 328.

25. M. Rolland, "Agrarian Problem in Mexico"; "News Notes and Personals," *Single Tax Review* 16 (November–December 1916): 378.

26. Important works on the history of Mexico's tax policies include Aboites and Jáuregui, *Penuria sin fin*; Haber, Razo, and Maurer, *Politics of Property Rights*; Kuntz Ficker, *Historia mínima de la economía mexicana*; B. Smith, "Building a State on the Cheap"; and Aboites Aguilar, *Excepciones y privilegios*.

27. George, *Our Land and Land Policy*, 69–70.

28. M. Rolland, "Trial of Socialism in Mexico," 85–86; "Rolland habla de la labor desarrollado en Yucatán por Salvador Alvarado," *La Prensa* (Mexico City), October 2, 1916, 6; "Yucatan's Revolution," *Jackson (MI) Citizen Press*, October 13, 1916, 17.

29. Mesa Andraca, *Con Salvador Alvarado*, 37.

30. Modesto Rolland, quoted in Francisco Pendás, letter to an unnamed woman, January 8, 1917, CEHM, fondo XXI, carp. 109, exp. 12479, f. 4–6; Chacón, "Salvador Alvarado and Agrarian Reform," 190.

31. Carranza quoted in Chacón, "Salvador Alvarado and Agrarian Reform," 190.

32. Hall, "Alvaro Obregón and the Politics of Mexican Land Reform," 213; Tannenbaum, *Mexican Agrarian Revolution*, 184.

33. Eiss, *In the Name of El Pueblo*, 117–18.

34. Mesa Andraca, *Con Salvador Alvarado*, 47.

35. "Many Mexican Problems Solved in Yucatán," *New York Times*, October 1, 1916, 3; M. Rolland, "Labor Law of Yucatan," 1–2; "Mexican Mob Burns Wilson in Effigy," *New York Tribune*, June 20, 1916, 2.

36. George, *Progress and Poverty*, 315.

37. See Alvarado, *La reconstrucción*, vol. 1.

38. Joseph, *Revolution from Without*, 146–47.

39. Joseph, *Revolution from Without*, 147.

40. Peniche Rivero, "Elvia Carrillo Puerto," 96–97. For more on women and the revolution in Yucatán, see S. Smith, "Removing the Yoke of Tradition"; and S. Smith, "Salvador Alvarado of Yucatán."

41. M. Rolland, "Women in Mexico," 29–31.

42. M. Rolland, "Women in Mexico," 29–31.

43. M. Rolland, "Women in Mexico," 29–31.

44. Joseph, *Revolution from Without*, 106.

45. Peniche Rivero, "Elvia Carrillo Puerto," 96–97.

46. Meyers, *Outside the Hacienda Walls*, 53–54.

47. "Many Mexican Problems Solved in Yucatán," *New York Times*, October 1, 1916, 3.

48. Joseph, *Revolution from Without*, 102.

49. Craib, *Cartographic Mexico*, 222–24.

50. M. Rolland, "Trial of Socialism in Mexico," 84.

51. Tannenbaum, *Peace by Revolution*, 117.

52. M. Rolland, *Problema de la Baja California*, 4–8; J. Castro, "Radiotelegraphy to Broadcasting," 335–65.

53. M. Rolland, *Problema de la Baja California*, 4–12.

54. "Una de las grandes obras de la revolución," *El Pueblo*, March 12, 1917, 10.

55. For more on the Esteban Cantú administration in Baja California, see Calvillo Velasco, "Indicios para descifrar la trayectoria política de Esteban Cantú"; and Vanderwood, *Satan's Playground*.

56. M. Rolland, *Problema de la Baja California*, 11–12; M. Rolland, *Informe sobre el Distrito Norte de la Baja California*; Schantz, "All Night at the Owl," 549–602.

5. War and Peace

1. "Only Bryan Lags in Move for Peace," *New York Times*, June 25, 1916, 2.
2. Rodgers, *Atlantic Crossings*, 267–90.
3. Eisenhower, *Intervention!*, 187–218.
4. Howe, *Campaigning in Mexico*, 3.
5. "Troops Ordered Rushed to Border," *Rock Island (IL) Argus*, June 23, 1916, 1; "Mexico Is Drilling Recruits While Guardsmen Organize," *Evening Star* (Washington DC), July 5, 1916, 2.
6. William Randolph Hearst quoted in Eisenhower, *Intervention!*, 292.
7. "Babricora [sic] Ranch Seized," *New York Times*, July 11, 1916, 6; "Mrs. Hearst Makes Protest," *New York Times*, July 14, 1916, 5.
8. William Randolph Hearst to E. T. O'Loughlin, July 5, 1916, reprinted in "Intervene in Mexico, Not to Make but to End War, Urges Mr. Hearst," *Brooklyn (NY) Eagle*, July 6, 1916, 8.
9. Stout, *Border Conflict*, 89.
10. Cook, *Crystal Eastman on Women and Revolution*, 6–16.
11. Daniels, *Always a Sister*, 1–3.
12. C. T. Hallivan, "La Unión Contra el Militarismo," 1916, CEHM, fondo XXI, carp. 84, exp. 9364, f. 1–5; "Pacifists Call Meeting at El Paso," *Los Angeles Times*, June 24, 1916, 2.
13. On Alvarado's funding, see M. C. Rolland to Venustiano Carranza, January 10, 1917, CEHM, fondo XXI, carp. 109, exp. 12479, f. 5.
14. Certificate of Incorporation of Columbus Publishing Company, Inc., May 31, 1916, CEHM, fondo XXI, carp. 109, exp. 12479, f. 1; Articles of agreement for Columbia Publishing Company, June 21, 1916, fondo XXI, carp. 109, exp. 12479, f. 1.
15. F. Pendás to V. Carranza, January 13, 1917, CEHM, fondo XXI, carp. 109, exp. 12479, f. 6; M. C. Rolland to Venustiano Carranza, January 16, 1917, CEHM, fondo XXI, carp. 109, exp. 12479, f. 5
16. Stein, "Lincoln Steffens," 197–98.
17. "Body to Try to Mediate Mexico Case," *Rock Island Argus*, June 23, 1916, 1; "Antimilitarism Body Would Prevent Conflict," *Colorado Springs Gazette*, June 24, 1916, 7; "Demonstration for Mexican Diplomat," *Sunday Times Advisor* (Trenton NJ), June 25, 1916, 1; "Rolland Off for Meeting to Avert War," *Fort Worth Star-Telegram*, June 25, 1916, 5; "Anti-War People Plan Conference," *San Jose Mercury Herald*, July 1, 1916, 2.
18. American Union against Militarism to Dr. Atl, June 22, 1916, CEHM, fondo XXI, carp. 85, exp. P514, f. 1–2.
19. Jordan, *Days of a Man*, 2:691.
20. "La actitud del pueblo americano contra los promotores de la guerra entre México y los Estado Unidos," *Acción Mundial* (Mexico City), July 21, 1916, 1.
21. Sáenz, *El símbolo y la acción*, 1–7, 204–60.
22. Kellogg, "New Era of Friendship for North America," 415.
23. Burns, *David Starr Jordan*, 1–25; Jordan, *Days of a Man*, 2:690.
24. "Remember Carrizal Says Mayor," *El Paso Herald*, July 6, 1916, 9.

25. "Jordan Not Driven Away from El Paso," *New York Times*, July 4, 1916, 18.

26. Kellogg, "New Era of Friendship for North America," 415.

27. "See a Catastrophe in War with Mexico," *New York Times*, July 5, 1916, 11.

28. "Anti-War People Plan Conference," *San Jose Mercury Herald*, July 1, 1916, 2.

29. "Wilson Stands Firm," *New York Times*, June 30, 1916, 3.

30. Jordan, *Days of a Man*, 2:698–99; Kellogg, "New Era of Friendship for North America," 415.

31. "See Flimsy Reason for War on Mexico," *Evening Star*, July 5, 1916, 2; "Commission Issues Statement," *Evening Star*, July 7, 1916, 9.

32. Mexican-American Peace Committee, "Mexican-American League," 1–7; "Working to Improve Mexican-American Relations," *Olympia (WA) Daily Recorder*, November 16, 1916, 4.

33. "Delegates to Make Tour," *New York Times*, July 9, 1916, 15; Britton, *Revolution and Ideology*, 119.

34. Mexican-American Peace Committee, "Mexican-American League," 1–7; "Working to Improve Mexican-American Relations," *Olympia Daily Recorder*, November 16, 1916, 4.

35. Stout, *Border Conflict*, 93–102.

36. Wilson quoted in Stout, *Border Conflict*, 91.

37. "Delegates to Make Tour," *New York Times*, July 9, 1916, 15. Also see Jordan, *Days of a Man*, 2:699–700.

38. Kellogg, "New Era of Friendship for North America," 416.

39. These speeches are in Rowe, *Purposes and Ideals of the Mexican Revolution*, 1–32.

40. M. Rolland, "Open Letter of Mr. M. C. Rolland to Mr. W. R. Hearst."

41. M. Rolland, "Open Letter of Mr. M. C. Rolland to Mr. W. R. Hearst."

42. M. Rolland, "Open Letter to the Honored President."

43. M. Rolland, "Open Letter to the Honored President."

44. M. Rolland, "Trial of Socialism in Mexico," 90.

45. "Se acelera la reconstrucción de México," *El Cosmopolitan* (Kansas City MO), November 18, 1916, 1; "Many Mexican Problems Solved in Yucatan," *New York Times*, October 1, 1916, 3.

46. "An Editorial Survey of the Horizon," *Herald Gospel of Liberty* 108, no. 27 (July 6, 1916): 838; "The Patch of Blue Sky in the Murky Mexican Sky," *Current Opinion* 61, no. 6 (December 1916): 370.

47. H.S.H, "All Exploitation in Mexico Is Not Vicious," *El Paso Herald*, July 21, 1916, 6.

48. W. B., "Do Interests Want Intervention?," *Forum*, September 1916, 280.

49. W. B., "Do Interests Want Intervention?," *Forum*, September 1916, 280.

50. Jordan, *Days of a Man*, 2:700.

51. Ricardo Flores Magón, "La agonía," *Regeneración* (Los Angeles), November 25, 1916, 1.

52. Jorge M. Rolland C., email to J. Justin Castro, February 23, 2013.

53. Modesto Rolland, quoted in Francisco Pendás, letter to an unnamed woman, January 8, 1917, CEHM, fondo XXI, carp. 109, exp. 12479, f. 4.

54. Francisco Pendás to an unnamed woman, January 8, 1917, CEHM, fondo XXI, carp. 109, exp. 12479, f. 4.

55. Modesto C. Rolland to Salvador Alvarado, October 4, 1916, AGEY, PE, G, caja 539, v. 204, exp. 18; Modesto C. Rolland, expenditure voucher, October 31, 1916, AGEY, PE, HP, caja 556, v. 201, exp. 18.

56. M. C. Rolland to Venustiano Carranza, January 16, 1917, CEHM, fondo XXI, carp. 109, exp. 12479, f. 2.

57. M. C. Rolland to Venustiano Carranza, January 16, 1917, CEHM, fondo XXI, carp. 109, exp. 12479, f. 2.

58. M. C. Rolland to Venustiano Carranza, January 10, 1917, CEHM, fondo XXI, carp. 109, exp. 12479, f. 5.

59. M. C. Rolland to Venustiano Carranza, January 10, 1917, CEHM, fondo XXI, carp. 109, exp. 12479, f. 5.

60. Informe de Correspondencia, August 8, 1916, CEHM, fondo XXI, carp. 92, exp. 10471, f. 1; "Información diaria extractada de la prensa," June 16, 1917, CEHM, fondo XXI, carp. 113, exp. 12976, f. 1.

61. Neumann Brothers to Venustiano Carranza, February 6, 1919, CEHM, fondo XXI, carp. 130, exp. 14841, f. 1.

62. M. C. Rolland, "El socialismo en Yucatán México," *El Gráfico* 2, no. 3 (January 1918): 422–23; Federico C. Howe, "El impuesto a la tierra y el trazado de las ciudades," *El Gráfico* 2, no. 3 (January 1918), 440–41.

63. Austin, "What the Mexican Conference Really Means," 1–6.

6. Transitions

1. V. Carranza, *Report by Venustiano Carranza*, 3–20. For more on the constitutional convention, see Ferrer Mendiolea, *Historia del Congreso Constituyente*; and Niemeyer, *Revolution at Querétaro*.

2. P. Smith, "Making of the Mexican Constitution," 186–90.

3. Knight, *Mexican Revolution*, 2:471–75.

4. Act verifying the election of Modesto C. Rolland and Manuel Villarino as representatives to the Constitutionalists' constitutional congress, 1917, AHPLM, RR, vol. 670 2/2, exp. s/n; "Los Diputados al Congreso Constituyente," *El Demócrata* (Mexico City), November 14, 1916, 1.

5. Joseph, *Revolution from Without*, 115.

6. M. C. Rolland, "Carta a mis ciudadanos," *La Información* (Mexico City), March 5, 1917, 4; Rolland, *Carta a mis conciudadanos*.

7. M. C. Rolland, "La neutralidad de México/The Neutrality of Mexico," *El Gráfico* 1, no. 7 (May 1917): 222–23; "Press Opinions," *The Public* 20, no. 599 (May 25, 1917): 4.

8. M. Rolland, *Carta a mis conciudadanos*, 32.

9. M. Rolland, *Carta a mis conciudadanos*, 32.

10. "Relación de Subarrendatarios del Ingeniero Modesto C. Rolland, en una Parte del Ateneo Peninsular," August 11, 1919, AGEY, PE, G, caja 703; C. R. Cardeña to the Oficial Mayor, Mérida, October 28, 1920, AGEY, PE, G, caja 74.

11. "Circulares," *Diario Oficial de Yucatán* (Mérida), April 3, 1918, 1943.

12. M. Rolland, *Carta a mis conciudadanos*, 29–31; M. Rolland, "Problema de la Baja California," 4–12; J. Castro, "Modesto C. Rolland and the Development of Baja California," 275–81.

13. "¿Conviene al país un puerto libre en Guaymas?," *El Universal* (Mexico City), July 4, 1920, 2; Rolland, *Reconstructive Policy in Mexico*, 10; Paoli, *Yucatán y los orígenes*, 109.

14. See Alvarado, *Salvador Alvarado*.

15. "No se puedo facilitar un crédito de $25,000.00 al comercio municipal," *El Pueblo*, June 24, 1918, 4.

16. Alvarado, *Carta al pueblo Yucatán*, 224.

17. M. C. Rolland to Carlos Castro Morales, November 6, 1918, AGEY, PE, caja 639.

18. Román Kalisch, "Desarrollo de tecnología constructiva," 69–75.

19. M. Rolland, "Agrarian Problem."

20. "La Ley de Reclamaciones," *El Heraldo de México* (Mexico City), September 1, 1919, 1, 8; "El libro del Gral. Salvador Alvarado," *El Heraldo de México*, September 5, 1919, 1; "La carta del Gral. Salvador Alvarado," *El Heraldo de México*, September 7, 1919, 1, 11; "Samuel Gompers se dirige a los obreros," *El Heraldo de México*, September 14, 1919, 1; "Esteban Cantú y su actuación administrativa," *El Heraldo de México*, September 15, 1919, 1, 11.

21. In particular see M. Rolland, *Carta a mis conciudadanos*.

22. "La conciencia nacl. ante el problema del petróleo," *El Heraldo de México*, September 1, 1919, 1, 12; "'El Heraldo de México' y el asunto del petróleo," *El Heraldo de México*, October 4, 1919, 1.

23. "Fue ofrecido un lunch-champagne a los periodistas," *El Pueblo*, April 27, 1919, 3.

24. "Brillante reunión," *El Demócrata*, December 31, 1919, 2.

25. "Balle en el Centro de Ingenieros," *El Universal*, August 11, 1919, 6; "Fiestas y recepciones," *El Universal*, January 20, 1920, 6.

26. Calvillo Velasco, "Prólogo," 9.

27. "Centro Nacional de Ingenieros," *Revista Mexicana de Ingeniería y Arquitectura* 4, no. 3 (March 15, 1926), 16; Baptista and Saldaña, "La participación política y reivindicación gremial del Centro de Ingenieros de México ante la construcción del Estado Mexicano en los años veinte," 1222.

28. "Balle en el Centro de Ingenieros," *El Universal*, August 11, 1919, 6.

29. "La última posada del Centro de Ingenieros," *El Universal*, December 25, 1919, 6.

30. Calvillo Velasco, "Prólogo," 19–20.

31. Calvillo Velasco, "Indicios para descifrar la trayectoria política de Esteban Cantú," 995–97.

32. "Estuvo en San Francisco el Lic. Millan y Alva," *Hispano America*, September 27, 1919, 2. Rolland's 1919 report was later republished in 1993 with an excellent introduction on Rolland and the context of the *Informe* by the historian Max Calvillo Velasco. See Calvillo Velasco, "Prólogo."

33. Schantz, "All Night at the Owl," 557.

34. Schantz, "All Night at the Owl," 558–60.

35. Schantz, "All Night at the Owl," 558–59; Núñez Tapia, "Aspectos del turismo en el Distrito Norte de Baja California," 59–60.

36. M. Rolland, *Informe sobre el Distrito Norte de la Baja California*, 160; "No Commissions to Pay in Cantu's Autocracy," *Los Angeles Times*, June 16, 1916, 13.

37. U.S. reporters often referred to Baja California during Cantú's reign as "Cantú's Kingdom" or the "Kingdom of Cantú." See Mary Brown Donoho, "Senor Cantu's Kingdom," *New York Times*, August 22, 1920, 7.

38. M. Rolland, *Informe sobre el Distrito Norte de la Baja California*, 35.

39. M. Rolland, *Informe sobre el Distrito Norte de la Baja California*, 159.

40. M. Rolland, *Informe sobre el Distrito Norte de la Baja California*, 161.

41. Ernesto Hidalgo, "Los presidenciables, sus amigos actuales y futuros," *El Universal*, March 30, 1919, 11; U.S. Congress, *Investigation of Mexican Affairs*, 2937.

42. T. Flores, *Contraespionaje político y sucesión presidencial*, 28–31.

43. Hall, *Álvaro Obregón*, 45, 59–62.

44. T. Flores, *Contraespionaje político y sucesión presidencial*, 28–31.

45. Castro, *Radio in Revolution*, 100. Also see Dulles, *Yesterday in Mexico*; Plasencia de la Parra, *Personajes y escenarios*; and Castro Martínez, *Adolfo de la Huerta*.

46. Velázquez Estrada, *Salvador Alvarado*, 28.

47. "Es muy importante el proyecto de los Puertos Libres," *Excélsior* (Mexico City), July 3, 1920, 10.

7. Opportunity, Defeat, and Death

1. "En honor de la prensa mexicana," *El Universal* (Mexico City), July 30, 1920, 8.

2. "Fue electo presidente provisional de la república el Sr. D. Adolfo de la Huerta," *El Heraldo de México* (Mexico City), May 25, 1920, 1.

3. Hall, *Álvaro Obregón*, 45–60.

4. Winberry, "Mexican Landbridge," 12–18; Garner, "Politics of National Development," 339–56.

5. Garner, *British Lions and Mexican Eagles*, 219.

6. E. Hidalgo, "¿Conviene al país un puerto libro en Guaymas?," *El Universal*, July 6, 1920, 2.

7. Thoman, *Free Ports*, xi–11.

8. There have been exceptions in which the whole city became a "free port."

9. "Es muy importante el proyecto de los puertos libres," *Excélsior* (Mexico City), July 3, 1920, 10; "El decreto sobre los puertos libres ya está terminado," *Excélsior*, July 14, 1920, 7; "Opinión en pro de los puertos libres," *El Universal*, August 23, 1920, 2.

10. M. Rolland, *Los puertos libres mexicanos*, 6–34; Jacoby, *Strange Career of William Ellis*, 180–81.

11. Julio Grandjean, "Que se entiende por puertos libres, cuáles son sus ventajas y cuáles pueden ser sus peligros para México," *El Universal*, August 5, 1920, 3; Fernando Leal Novelo, "Sobre los puertos libres," *El Universal*, August 8, 1920, 5; "Como juzgan la creación de los puertos libres y los industriales," *El Universal*, August 11, 1920, 2.

12. "Habrá por fin puertos libres," *El Universal*, August 4, 1920, 1.

13. "Una conferencia del ingeniero M. Roland [sic]," *Excélsior*, July 16, 1920, 7; "Los puertos libres son ventajosos," *Excélsior*, July 28, 1920, 1.

14. "Puertos libres en México," *El Universal*, November 26, 1920, 1.

15. "Proyectos de la comisión de puertos libres," *El Universal*, November 21, 1921, 16.

16. Tannenbaum, *Mexican Agrarian Revolution*, 225.

17. "Distribución de expedientes en la Comisión Agraria," *El Universal*, July 22, 1920, 8.

18. Hall, "Alvaro Obregón and the Politics of Mexican Land Reform," 215–30; Shadle, *Andrés Molina Enríquez*, 76.

19. "Asociación para estudiar los problemas nacionales," *El Universal*, March 10, 1921, 3.

20. "Asociación para estudiar los problemas nacionales," *El Universal*, March 10, 1921, 3; "Una conferencia del Ing. Modesto Rolland," *Excélsior*, May 27, 1921, 3; "Notas sociales y personales," *Excélsior*, July 15, 1920, 3.

21. M. Rolland, *El desastre municipal*, 17–18.

22. M. Rolland, *El desastre municipal*, 17.

23. B. Smith, "Building a State on the Cheap," 257.

24. M. Rolland, *El desastre municipal*, 27–95.

25. M. Rolland, "Agrarian Problem."

26. "Proyecto para establecer servicio civil," *El Universal*, November 29, 1920, 1.

27. Hall, "Alvaro Obregón and the Politics of Mexican Land Reform," 216–17; Shadle, *Andrés Molina Enríquez*, 76; Sanderson, *Agrarian Populism*, 78–83.

28. "Estuvo movida la sesión del Congreso Nacional Agronómico," *Excélsior*, September 11, 1921, 1, 7; "La Comisión N. Agraria pretende servirse del Congreso Agronómica," *Excélsior*, September 13, 1921, 1.

29. "Una proposición para suprimir los ayuntamientos en el país," *Excélsior*, April 25, 1922, 6.

30. Gabriel Oropeza, "Anales de la primera Convención Nacional de Ingenieros," *El Universal*, November 1922, sec. 3, 4.

31. Shadle, *Andrés Molina Enríquez*, 88–89.

32. Tannenbaum, *Mexican Agrarian Revolution*, 245.

33. Rolland, "Agrarian Problem"; "Comité Ejecutivo de Administración de ejidos," *El Universal*, September 25, 1921; Sanderson, *Agrarian Populism*, 79.

34. Rolland, "Agrarian Problem"; Tannenbaum, *Mexican Agrarian Revolution*, 239. The word *ejido* was often spelled *egido* in the 1920s.

35. Modesto C. Rolland to Álvaro Obregón, February 6, 1922, AGN, RPR, Obregón-Calles, caja 194, exp. 609-R-5.

36. Álvaro Obregón to Modesto C. Rolland, February 7, 1922, AGN, RPR, Obregón-Calles, caja 194, exp. 609-R-5.

37. Hall, "Alvaro Obregón and the Politics of Mexican Land Reform," 219–37.

38. For more the development of radio in Mexico, see Castro, *Radio in Revolution*; Hayes, *Radio Nation*; Mejía Barquera, *La industria de la radio y televisión*; and Medina Ávila and Vargas Arana, *Nuestra es la voz*. Some of this section on radio comes from J. Castro, *Radio in Revolution*.

39. J. Castro, *Radio in Revolution*, 104–38.

40. Ornelas Herrera, "Radio y cotidianidad en México," 145; Albarrán, *Seen and Heard*, 133.

41. Mejía Barquera, *La industria de la radio y televisión*, 35.

42. "Los permisos para las estaciones de radiotelefonía," *El Universal*, September 1, 1923, 3.

43. "Una interesante conferencia sobre radiotelefonía," *El Universal*, March 4, 1923, 4.

44. "Piden se anule el reglamento para el radio," *Excélsior*, June 14, 1923, sec. 2, 1.

45. "La feria del radio fue inaugurada por el Presidente de la República, ayer," *Excélsior*, June 17, 1923, sec. 2, 1; "La próxima feria de radio en la capital," *Excélsior*, May 27, 1923, sec. 3, 9.

46. "Mexico Will Open 2 New Free Ports," *New York Times*, July 31, 1923, 33.

47. "Mexico Will Open 2 New Free Ports," *New York Times*, July 31, 1923, 33.

48. Free Ports Executive Board, *Mexican Free Ports*, 77.

49. "Mexico Will Open 2 New Free Ports," *New York Times*, July 31, 1923, 33.

50. Jorge M. Rolland C., email to J. Justin Castro, October 23, 2012.

51. Jorge M. Rolland C., email to J. Justin Castro, May 2, 2016.

52. Deanna Catherine Wicks, email to J. Justin Castro, March 5, 2016; W. H. Ellis to Charles Beecher Warren and John Barton Payne, July 13, 1923, NARA, RG 59, 812.1561/10.

53. "La imprudencia del chofer causó la muerte de la Sra. Virginia Garza de Rolland," *El Mundo* (Mexico City), December 6, 1–2.

54. Hall, "Alvaro Obregón and the Politics of Mexican Land Reform," 229–31.

55. "Rebels Say They Hold the Oil Fields," *New York Times*, February 17, 1924, 6.

56. Deanna Catherine Wicks, Facebook message to J. Justin Castro, October 29, 2016.

57. "Evacuation of Puebla Reported," *New York Times*, December 14, 1923, 2.

58. Dulles, *Yesterday in Mexico*; Plasencia de la Parra, *Personajes y escenarios*; Castro Martínez, *Adolfo de la Huerta*.

59. "Formal batida contra los aparatos radiotelefónicos," *El Universal Gráfico* (Mexico City), January 16, 1924, 2; "El acuerdo del General Gómez," *Excélsior*, January 16, 1924, 1.

60. "3 Mexican Ports Opened for Business," *New York Times*, July 22, 1924, 31.

61. "3 Mexican Ports Opened for Business," *New York Times*, July 22, 1924, 31.

62. M. Rolland, "Free Ports and the Interoceanic Traffic," 4, 15.

63. M. Rolland and Irigoyen, *Justificación*, 10.

64. "Serios obstáculos se presentan al comercio mexicano," *La Prensa* (San Antonio TX), November 9, 1939, 4.

8. A Stadium for Stridentopolis

1. Manuel Maples Arce, "TSH (El poema de la radiofonía)," *El Universal Ilustrado* (Mexico City), April 5, 1923, 19. TSH is an acronym for *telegrafía sin hilos*, which translates to radiotelegraphy/radiotelephony without wires, or wireless. I have provided this prose translation, but for a slightly different take in English, see Gallo, *Mexican Modernity*, 127.

2. See his memoirs: Maples Arce, *Soberana juventud*.

3. J. Castro, "Sounding the Mexican Nation," 4–8.

4. Flores Ayala, "Adalberto Tejada," 301–12.

5. Maples Arce, *Soberana juventud*, 183–91.

6. Hernández Palacios, *Xalapa de mis recuerdos*, 13.

7. Klich, "Estridentópolis," 117; Rashkin, *Stridentist Movement*, 170; Hernández Palacios, *Xalapa de mis recuerdos*, 17.

8. "Calles Abolishes Mexican Free Ports," *Oakland (CA) Tribune*, April 30, 1925, 1.

9. Mendoza, "Estadio Xalapeño Heriberto Jara Corona," 342–44.

10. Heriberto Jara quoted in Zapata Vela, *Conversaciones con Heriberto Jara*, 57–58; J. M. Rolland, "Estadio Xalapeño," 72; Mendoza, "Estadio Xalapeño Heriberto Jara Corona," 336–37.

11. Zapata Vela, *Conversaciones con Heriberto Jara*, 57–58.

12. Klich, "Estridentópolis," 117–18; Gallo, *Mexican Modernity*, 183, 205–7. The stadium was officially designated the Estadio Heriberto Jara Corona, but it has been commonly referred to as the Estadio Xalapeño. In the 1920s most people would have used the spelling Jalapeño and Jalapa. I changed the *J* to *X* to reflect the current spelling.

13. M. Rolland, *Jalapa-Enríquez*, 1.

14. "Estadio Xalapeño, 90 años de resistir el tiempo," *El Heraldo de Xalapa*, August 17, 2015, 4.

15. Jorge M. Rolland C., email to J. Justin Castro, April 11, 2016.

16. Hernández Palacios, *Xalapa de mis recuerdos*, 13.

17. Hernández Palacios, *Xalapa de mis recuerdos*, 35.

18. Wood, "Adalberto Tejada of Veracruz," 77–79.

19. M. Rolland, *Jalapa-Enríquez*, 1–2.

20. Pollock and Morgan, *Modern Cities*, 1–12 (quotes, 8, 9, 2, 1, respectively).

21. Jara vaguely discusses his ideological beliefs in Zapata Vela, *Conversaciones con Heriberto Jara*, 173–75.

22. P. Hall, *Cities of Tomorrow*, 87.

23. Howard, *Garden Cities of To-morrow*, 140. This quote is also discussed in P. Hall, *Cities of Tomorrow*, 94.

24. P. Hall, *Cities of Tomorrow*, 96.

25. There is little to no literature on the garden city movement's influence in Latin America, but there is already an impressive literature on the movement's influence in Europe and the United States. See P. Hall, *Cities of Tomorrow*; Marx, *Machine in the Garden*; Gillette, *Civitas by Design*; Schaffer, *Garden Cities for America*; Sutcliffe, *Rise of Modern Urban Planning, 1800–1914*; and Sutcliffe, *Metropolis, 1890–1940*.

26. M. Rolland, *Jalapa-Enríquez*, 3.

27. Rashkin, *Stridentist Movement*, 172.

28. "Athens 1896," Olympic Games, accessed April 22, 2016, http://www.olympic.org/athens-1896-summer-olympics.

29. "Paris 1924," Olympic Games, accessed April 22, 2016, http://www.olympic.org/paris-1924-summer-olympics; Celestino Herrera Frimont, "La revolución y la educación física." *Horizonte* 1, no. 8 (November 1926): 13–14; Gallo, *Mexican Modernity*, 202.

30. Trumpbour, *New Cathedrals*, 11.

31. Gallo, *Mexican Modernity*, 204.

32. Gallo, *Mexican Modernity*, 201–26.

33. J. Rolland, "Estadio Xalapeño," 72.

34. Sánchez Fogarty quoted in Gallo, *Mexican Modernity*, 184.

35. "Estadio Xalapeño," *El Heraldo de Xalapa*, August 17, 2015, 4; J. Rolland, "Estadio Xalapeño," 74.

36. M. Rolland, *Jalapa-Enríquez*, 3.

37. M. Rolland, *Jalapa-Enríquez*, 3, 10.

38. Hernández Palacios, *Xalapa de mis recuerdos*, 26.

39. J. Rolland, "Estadio Xalapeño," 71–74.

40. Modesto C. Rolland to Heriberto Jara, September 30, 1925, JMR.

41. Klich, "Estridentópolis," 118.

42. Klich, "Estridentópolis," 119.

43. Manuel Maples Arce, "La Convención Azucarera," *Horizonte* 1, no. 4 (July 1926): 17–19.

44. Mendoza, "Heriberto Jara Corona," 342–45.

45. "El Sr. Presidente inauguró ayer el estadio de Jalapa," *Excélsior* (Mexico City), September 21, 1925, FAPECFT, FJA, exp. 165, inv. 476, leg. 6/12.

46. Modesto C. Rolland quoted in J. M. Rolland, "Estadio Xalapeño," 74.

47. "El Sr. Presidente inauguró ayer el estadio de Jalapa," *Excélsior*, September 21, 1925, FAPECFT, FJA, exp. 165, inv. 476, leg. 6/12; Klich, "Estridentópolis," 119.

48. Hernández Palacios, *Xalapa de mis recuerdos*, 25–27.

49. Klich, "Estridentópolis," 119.

50. Henry George, "Vena a nos él tu reino," *Horizonte* 1, no. 2 (May 1926): 11–14; Leo Tolstoy, "La única solución posible de la cuestión agraria," *Horizonte* 1, no. 2 (May 1926): 19–22.

51. Maples Arce quoted in J. M. Rolland, "Estadio Xalapeño," 75.

52. Manuel Maples Arce, "Nuevas ideas—la estética del sidero-cemento," *Horizonte* 1, no. 3 (June 1926): 9–11.

53. Germán List Arzubide, "Construid un estadio, mensaje a la provincial," *Horizonte* 1, no. 1 (April 1926): 9–11.

54. Rashkin, *Stridentist Movement*, 172.

55. Rashkin, "La arqueología de Estridentópolis," xxvii–xxxi.

56. Maples Arce, *Soberana juventud*, 206–10.

57. Berman, *All That Is Solid Melts into Air*, 244.

9. Mr. Bothersome

1. For a solid discussion of the increasing frustration of technocratic reformers, see Gauss, *Made in Mexico*, 1–45.

2. "Un magnifico concierto de radio para esta noche por la estación 'Excélsior,'" *Excélsior* (Mexico City), September 28, 1925, 4; Hershfield, *Imagining la Chica Moderna*, 60.

3. For a work on foreign intellectuals in Mexico, see Delpar, *Enormous Vogue of Things Mexican*; Tenorio-Trillo, "Cosmopolitan Mexican Summer," 224–42; La Botz, "American 'Slackers' in the Mexican Revolution," 563–90.

4. Castillo Peraza, *Manuel Gómez Morín*, 16. Also see Lara G., *Manuel Gómez Morín*.

5. Buchenau, *Plutarco Elías Calles*, 119.

6. Maurer, *Power and the Money*, 177–94; Haber, *Industry and Underdevelopment*, 125–31.

7. Haber, *Industry and Underdevelopment*, 150–56; Buchenau, *Plutarco Elías Calles*, 206–9; Horn, "U.S. Diplomacy," 31–45.

8. "President Calles States the Case for the Laws Mexico Is Enforcing," *New York Times*, August 1, 1926, 1.

9. Fallaw, *Religion and State Formation in Postrevolutionary Mexico*, 15; J. Meyer, *La Cristiada*, 1:7–13; Bailey, *¡Viva Cristo Rey!*, 48–75.

10. Weis, "Revolution on Trial," 320–53; Buchenau, *Plutarco Elías Calles*, 143.

11. Calles, *Pensamiento político y social*, 291.

12. Krauze, *Mexico*, 428.

13. Buchenau, *Plutarco Elías Calles*, 144–54.

14. This accusation was unfounded.

15. Dulles, *Yesterday in Mexico*, 460–61; Buchenau, *Plutarco Elías Calles*, 148–53.

16. "Lower California and Tehuantepec," *New York Times*, December 26, 1930, 12.

17. The word *molesto* in Spanish has a meaning that is closer to "irksome" or "bothersome." It does not have the same sexual connotations as the verb "molest" often does in English. Rolland had had this nickname since 1916, if not before. See, Mesa Andraca, *Con Salvador Alvarado*, 69.

18. Deanna Wicks, email to J. Justin Castro, April 11, 2016.

19. Deanna Wicks, email to J. Justin Castro and Jorge M. Rolland C., May 4, 2016.

20. Modesto Rolland, "Vida y hechos del Ing. M. C. Rolland," unpublished document, 1964, JMR; Jorge M. Rolland C., email to J. Justin Castro, May 3, 2016.

21. M. Rolland, "Vida y hechos del Ing. M. C. Rolland."
22. Jorge M. Rolland C., email to J. Justin Castro, May 3, 2016; "Las Minas," Source Exploration Corp., accessed May 24, 2016, http://www.sourceexploration.com/section.asp?catid=2364&subid=4894; J. Rolland, *Modesto C. Rolland*, 146–47.
23. *Aero-Motor México*, pamphlet, 1932, 1. I obtained a copy of this pamphlet from Jorge M. Rolland.
24. *Aero-Motor México*, 1–3.
25. *Aero-Motor México*, 4.
26. "La conquista del desierto," *El Nacional* (Mexico City), April 24, 1932, sec. 2, 4.
27. M. C. Rolland to P. E. Calles and Guerro Anzures, January 22, 1932, FAPECFT, APEC, exp. 238, inv. 5059, leg. 1.
28. Patente 18595, inventor Modesto C. Rolland, October 27, 1919, AGN, P, exp. 49, leg. 15.
29. L. O. Madero, "Un nuevo aero-motor de gran utilidad ha sido inventado aquí," *El Nacional*, March 10, 1932, 1–2.
30. M. Rolland, "Agrarian Problem in Mexico."
31. M. Rolland, "Agrarian Problem in Mexico."
32. M. Rolland, "Agrarian Problem in Mexico."
33. Gauss, *Made in Mexico*, 95.
34. Gauss, *Made in Mexico*, 32–42.
35. M. Rolland, "Agrarian Problem in Mexico."
36. M. Rolland, "Agrarian Problem in Mexico."
37. M. Rolland, *Comunismo o liberalismo?*, 5–7. Quotations come from a version of the essay published as a stand-alone second edition.
38. M. Rolland, *Comunismo o liberalismo?*, 7–10.
39. Deanna Catherine Wicks, email to J. Justin Castro, October 15, 2016; Jorge M. Rolland, email to J. Justin Castro, October 31, 2016.
40. Cockcroft, *Mexico's Revolution*, 63.
41. Carr, *Marxism and Communism*, 43–45; Martínez Verdugo, *Historia del comunismo*, 109–20.
42. Dulles, *Yesterday in Mexico*, 480.
43. Spencer and Stoller, "Radical Mexico," 57–70.
44. J. Rolland, "Un constructor del México moderno," 82.
45. M. Rolland, *Justificación*, 2; J. Rolland, "Life of Modesto C. Rolland: Resume in English," *Modesto C. Rolland* (blog), July 25, 2011, http://modestoroland.blogspot.com/2011/07/life-of-modesto-c-rolland-resume-in.html; Jorge M. Rolland C., email to J. Justin Castro, April 25, 2016. For more on the Ferrocarril del Sureste, see Balcázar Antonio, *Tabasco a dos tiempos*.
46. Francis M. Withey, "The Reconstruction of the Tehuantepec Railroad and Related Port Facilities in Salina Cruz," confidential memorandum report for the American embassy, March 15, 1942, NARA, RG 84, 870/690.
47. Parra, *Writing Pancho Villa's Revolution*, 18; Dulles, *Yesterday in Mexico*, 493.
48. Gauss, *Made in Mexico*, 33–34.

10. The Undersecretary

1. Diego Rivera, Frida Kahlo, et al., quoted in Tibol, *Frida Kahlo*, 103–7 (quote, 105–6). This is the only reference I have found that mentions Rolland inventing a tortilla machine or hanging baskets for potatoes.
2. Tibol, *Frida Kahlo*, 103–7.
3. Lázaro Cárdenas, *El sr. Gral. de División Lázaro Cárdenas: Candidato a la presidencia de la república de México por el P.N.R., hace profesión de fe cooperativista* (Mexico City: Liga Nacional de Acción Cooperativa, 1934), 1–4, copy in FAPECFT, FP, exp. 5, inv. 803, leg. 1.
4. Grunstein Dickter, "In the Shadow of Oil," 14.
5. Cárdenas, "Al Gral. Lázaro Cárdenas," 25.
6. Grubb, "Political Situation in Mexico," 686.
7. Francisco J. Mújica to Lázaro Cárdenas, July 25, 1938, AGN, RPR, Lázaro Cárdenas, caja 0650, exp. 523/83; "El Ferrocarril del Sureste estará terminado en 1940," *El Porvenir* (Monterrey), August 8, 1938, 2. Cárdenas reinstated the free ports in April 1939.
8. Grunstein Dickter, "In the Shadow of Oil," 6.
9. Grunstein Dickter, "In the Shadow of Oil," 15–16.
10. Boyer, "Old Loves, New Loyalties," 435.
11. "El Ferrocarril del Sureste estará terminado en 1940," *El Porvenir*, August 8, 1938, 2.
12. Secretaria de Comunicaciones y Obras Públicas, *Ferrocarril del Sureste*, 40–197; Balcázar Antonio, *Tabasco a dos tiempos*, 113–27.
13. "Se nombró nuevo subsecretario de comunicaciones," *El Informador* (Guadalajara), October 6, 1938, 1.
14. "Dos edificios inaugurados en el P. Aereo," *El Informador*, November 28, 1938, 1.
15. "Decreto," *Diario Oficial del Gobierno del Estado de Yucatán*, September 17, 1940, 1–2.
16. Modesto C. Rolland to the Presidente de la República, December 13, 1939, AGN, RPR, Cárdenas, caja 575, exp., 506.11/77.
17. "Secretaría de la Economía Nacional," *Periódico Oficial de Nayarit*, September 11, 1940, 1–3.
18. Diego River, Frida Kahlo, et al., quoted in Tibol, *Frida Kahlo*, 106.
19. Niblo, *War, Diplomacy, and Development*, 69.
20. Schuler, *Mexico between Hitler and Roosevelt*, 53–55, 103–6.
21. Niblo, *War, Diplomacy, and Development*, 68; Paz, *Strategy, Security, and Spies*, 36–37.
22. "Jap-Controlled Oil Company to Exploit Mexican Field," *Corsicana (TX) Daily Sun*, 1; "Japs Want Mexican Oil," *Paris (TX) News*, October 16, 1940, 5; Paz, *Strategy, Security, and Spies*, 39.
23. Paz, *Strategy, Security, and Spies*, 44, 150.
24. William B. Richardson quoted in Niblo, *War, Diplomacy, and Development*, 68.

25. Wasserman, *Pesos and Politics*, 5–28.

26. "Mexico Defends Jap Oil Deal," *Daily Record* (Mexico City), October 19, 1940, 1; Schuler, *Mexico between Hitler and Roosevelt*, 104.

27. "Mexico Defends Jap Oil Deal," *Daily Record*, October 19, 1940, 1; Pierre de L. Boal to the secretary of state, October 17, 1939, NARA, RG 59, 812.00/30843.

28. Josephus Daniels to the secretary of state, November 13, 1940, NARA, RG 59, 812.6363/7136.

29. Josephus Daniels to the secretary of state, October 22, 1940, NARA, RG 59, 812.6363/7146; Paz, *Strategy, Security, and Spies*, 41–44.

30. Josephus Daniels to the secretary of state, October 25, 1940, NARA, RG 59, 812.6363/7151; "Fue cancelada la concesión petrolera a una compañía en la que figuraban japoneses," *Excélsior* (Mexico City), October 25, 1940, copy in NARA, RG 59, 812.6363/7151.

31. Paz, *Strategy, Security, and Spies*, 39; Schuler, *Mexico between Hitler and Roosevelt*, 104.

32. "Serios Obstáculos de presentan al comercio mexicano," *La Prensa* (San Antonio TX), November 9, 1939, 4; "México necesita puertos y barcos," *La Prensa*, September 26, 1939, 5.

33. Secretaria de Comunicaciones y Obras Públicas, *Ferrocarril del Sureste*, 197.

34. "Cardenas Men Named to Cabinet," *Daily News* (location not identified), January 12, 1939, found in NARA, RG 59, 812.911.

35. Modesto C. Rolland to Melquiades Angulo, February 20, 1939, AGN, RPR, Cárdenas, caja 1354, exp. 711/417.

36. Modesto C. Rolland to Melquiades Angulo, February 20, 1939, AGN, RPR, Cárdenas, caja 1354, exp. 711/417.

37. "Mexican Light and Power Firm Expands," *Valley Morning Star* (Harlingen TX), September 1, 1940, 2.

38. Modesto C. Rolland to Lázaro Cárdenas, August 23, 1939, AGN, RPR, caja 0670, exp. 527.1/27.

39. Cruz, *Vida y obra de Pastor Rouaix*, 217–19.

40. Francisco J. Mújica [or Múgica] to Lázaro Cárdenas, July 25, 1938, AGN, RPR, Lázaro Cárdenas, caja 0650, exp. 523/83; Francisco J. Mújica to Lázaro Cárdenas, July 27, 1938, AGN, RPR, Lázaro Cárdenas, caja 0650, exp. 523/83.

41. Agustín Leñero to Modesto C. Rolland, August 16, 1939, AGN, RPR, Lázaro Cárdenas, caja, 0621, exp. 513.3/22.

42. "El presidente navega hacia Bahía Magdalena," *El Nacional*, July 12, 1939.

43. This free zone was established in 1937.

44. Modesto C. Rolland, "Observaciones realizadas en la jira por la Baja California en compañía c. Presidente de la República," January 14, 1940, *Novedades*, 1–7, typewritten copy in AHPLM, RR, doc. 14, vol. 928½, exp. s/n; Castro, "Modesto C. Rolland," 281–83.

45. Modesto C. Rolland, "Observaciones realizadas en la jira por la Baja California en compañía c. Presidente de la República," January 14, 1940, *Novedades*, 1–7, typewritten copy in AHPLM, RR, doc. 14, vol. 928½, exp. s/n; Castro, "Modesto C. Rolland," 281–83.

46. Modesto C. Rolland, "Observaciones realizadas en la jira por la Baja California en compañía c. Presidente de la República," January 14, 1940, *Novedades*, 1–7, typewritten copy in AHPLM, RR, doc. 14, vol. 928½, exp. s/n.

47. Modesto C. Rolland, "Observaciones realizadas en la jira por la Baja California en compañía c. Presidente de la República," January 14, 1940, *Novedades*, 6, typewritten copy in AHPLM, RR, doc. 14, vol. 928½, exp. s/n.

48. Modesto C. Rolland to the Presidente de la República, December 13, 1939, AGN, RPR, Cárdenas, caja 575, exp., 506.11/77.

49. Miguel A. de Quevado to Agustín Leñero, April 13, 1939, AGN, RPR, Cárdenas, caja 575, exp. 506.11/77; Modesto Rolland to Lázaro Cárdenas, December 6, 1939, AGN, RPR, Cárdenas, caja 489, exp. 437.1/413.

50. Modesto C. Rolland, "Observaciones realizadas en la jira por la Baja California en compañía c. Presidente de la República," January 14, 1940, *Novedades*, 1–7, typewritten copy in AHPLM, RR, doc. 14, vol. 928½, exp. s/n.

51. L. Cárdenas to Modesto C. Rolland, December 17, 1939, AGN, RPR, Cárdenas, caja 575, exp., 506.11/77; Modesto C. Rolland to the Presidente de la República, December 13, 1939, AGN, RPR, Cárdenas, caja 575, exp. 506.11/77.

52. Modesto C. Rolland to the Presidente de la República, December 7, 1939, AGN, RPR, Cárdenas, caja 575, exp. 506.11/77.

53. Rolland, "Free Ports and the Interoceanic Traffic," 22.

54. "Free Trade with Americans Urged," *Los Angeles Times*, June 23, 1940, 20.

55. "Isthmus Line Is Furthered," *Arizona Republic*, February 4, 1940, 8.

56. Rolland and Irigoyen, *Justificación*, 1–14; Rolland, "Free Ports and the Interoceanic Traffic," 1–22.

57. Jacoby, *Strange Career of William Ellis*, 52.

11. Going Big

1. "Mexico Opens Inquiry," *New York Times*, December 19, 1940, 6; "Fue consignado el ex-Secretario de Economía Modesto Rolland," *La Prensa* (San Antonio TX), December 23, 1940, 8.

2. "Mexico Opens Inquiry," *New York Times*, December 19, 1940, 6.

3. "La corrupción del régimen pasado tenía tremendas proporciones," *La Prensa*, December 22, 1940, 1.

4. "La corrupción del régimen pasado tenía tremendas proporciones," *La Prensa*, December 22, 1940, 1.

5. "Se defiende el Ingeniero Don Modesto Roland [sic]," *Excélsior* (Mexico City), December 19, 1940, 1, 7.

6. Rankin, ¡*México, la Patria!*, 2–12, 281–91.

7. Susan Gauss provides a straightforward definition of ISI: "ISI is a type of forced industrialization aimed at replacing imports and characterized by the use of protection to encourage the domestic production of manufactured goods." Gauss, *Made in Mexico*, 5.

8. Relevant consular reports in NARA, RG 84; "U.S. Consulate Reopened," *New York Times*, September 30, 1940, 10.

9. "20 años de intense estudio," *La Prensa*, June 9, 1941, sec. 2, 1.

10. "Istmo de Tehuantepec," *Excélsior*, March 21, 1968, 4A.

11. Francis M. Withey, "The Reconstruction of the Tehuantepec Railroad and Related Port Facilities in Salina Cruz," confidential memorandum report for the American embassy, March 15, 1942, NARA, RG 84, 870/690.

12. Francis W. Withey to the secretary of state, January 25, 1943, NARA, RG 59, 812.00, Oaxaca/9.

13. Francis W. Withey to the secretary of state, February 22, 1943, March 15, 1943, NARA, RG 59, 812.00, Oaxaca/9; Francis W. Withey to the secretary of state, April 5, 1943, NARA, RG 59, 812.00, Oaxaca/12.

14. Rubén Calatayud, "Modesto C. Rolland," *El Mundo de Córdoba*, December 26, 2014, http://www.elmundodecordoba.com/index.php?option=com_content&view=article&id=3455681:col1&catid=218:columnas&Itemid=96.

15. Deanna Wicks, email to J. Justin Castro, October 16, 2016.

16. Modesto C. Rolland to Jorge Rolland, November 26, 1943, JMR.

17. Jorge M. Rolland, email to J. Justin Castro, June 30, 2016.

18. Modesto C. Rolland to Jorge Rolland, March 23, 1943, November 26, 1943, JMR.

19. For more on the Mexican Revolution in Chiapas, see Lewis, *Ambivalent Revolution*.

20. Boyer, "Cycles of Mexican Environmental History," 11.

21. Modesto C. Rolland to J. Jesús González Gallo, September 5, 1944, AGN, RPR, Ávila Camacho, caja 0523, exp. 195.1/5.

22. Gilderhus, *Second Century*, 98; "Mexicans Seeking Oil in Three New Areas," *New York Times*, January 2, 1945, 20.

23. "Mexican Oil Discussed," *New York Times*, March 23, 1942, 6; "Hay acumulación de carga en todo el país," *La Prensa*, June 3, 1943, 8.

24. For more on how World War II and governments during the 1940s clamped down on railroad worker activism see Alegre, *Railroad Radicals*, 38–44.

25. Manuel Ávila Camacho to the Gerente General de los Ferrocarriles Nacionales de México, July 10, 1943, AGN, RPR, Ávila Camacho, caja 0015, exp. 110.1/68.

26. Presidencial acuerdo, January 21, 1943; Modesto C. Rolland to the Oficial Mayor, April 9, 1943; Heriberto Jara to Manuel Ávila Camacho, July 22, 1943, all in AGN, RPR, Ávila Camacho, caja 0015, exp. 110.1/90.

27. Modesto C. Rolland to Manuel Ávila Camacho, July 27, 1943; Modesto C. Rolland to Heriberto Jara, July 27, 1943, both in AGN, RPR, Ávila Camacho, caja 0015, exp. 110.1/90.

28. Waldo Romo Castro to Manuel Ávila Camacho, August 2, 1943, AGN, RPR, Ávila Camacho, caja 0015, exp. 110.1/90.

29. Adolfo Ruiz Cortines to Manuel Ávila Camacho, December 22, 1945, AGN, RPR, Ávila Camacho, caja 0015, exp. 110.1/90.

30. Emilio Barragan and Nicolas López to Manuel Ávila Camacho, March 29, 1946, AGN, RPR, caja 0523, exp. 495.1/5.

31. Gilderhus, *Second Century*, 98–102.

32. J. Rolland, "Un constructor de México moderno," 83; J. Rolland, "Modesto C. Rolland: Vida y obra," 29; "Ciudad de Los Deportes," *Excélsior*, August 5, 1944, JMR.

33. "Ciudad de Los Deportes," *Excélsior*, August 5, 1944, JMR; "Ciudad de Los Deportes," *Excélsior*, August 9, 1944, JMR.

34. "Ciudad de Los Deportes," *Excélsior*, July 8, 1944, JMR.

35. J. Rolland, "Modesto C. Rolland: Vida y obra," 29.

36. "Cartel español y mexicano en al aniversario 66 de Plaza México," *Excélsior*, February 4, 2012, http://www.excelsior.com.mx/2012/02/04/adrenalina/807838.

37. J. Rolland, "La monumental Plaza de Toros México," 66–67; Rafael Morales, "La monumental," *Querétaro*, March 1996, 60.

38. Milton Bracker, "Bull Flees, Matador Turns Sword on Fan, So 50,000 Rioting Mexicans Wreck Stadium," *New York Times*, January 21, 1947, 11.

39. Benjamin, "Rebuilding the Nation," 500.

40. Ayala Espino, *Estado y desarrollo*, 288; Babb, *Managing Mexico*, 78–80.

41. Chavez quoted in "Mexico Suggested as a New Canal Site," *New York Times*, December 25, 1947, 14.

42. M. Rolland, *Overland Ship Transportation*, 4.

43. "Death of James B. Eads," *New York Times*, March 11, 1887, 5. There is a biography of Eads that discusses his project in some detail. See Dorsey, *Road to the Sea*.

44. "Death of James B. Eads," *New York Times*, March 11, 1887, 5; "Capt. Eads Ship Railroad Project," *New York Times*, November 11, 1880, 3.

45. "Mr. Eads' Great Ship Railway for the American Isthmus," *Scientific American* 43, no. 20 (November 13, 1880), 303; M. Rolland, *Transportation of Boats*, 7–8.

46. See R. Alexander, *Sons of the Mexican Revolution*.

47. R. Alexander, *Sons of the Mexican Revolution*, 1–15.

48. Niblo, *Mexico in the 1940s*, 169.

49. M. Rolland, *Transportation of Boats*, 7–18.

50. M. Rolland, *Transportation of Boats*, 20–30.

51. "Mexico Suggests 'Land Canal' to Supplement Panama Canal," *Blytheville (AR) Courier News*, March 5, 1948, 2.

52. "Proyecto para transportar barcos por ferrocarril," *La Prensa*, October 30, 1948, 2.

53. William Price, "Mexican Motoring," *New York Times*, October 24, 1948, X15.

54. "Mexico Gets $8 Million Loan from 2 U.S. Banks to Complete Highway," *Wall Street Journal*, October 6, 1948, 8.

55. "'Canal' on Rails May Link Oceans," *Los Angeles Times*, January 24, 1949, 10.

56. "'Canal' on Rails May Link Oceans," *Los Angeles Times*, January 24, 1949, 10.

57. "Mexico Studies Railway for Ships," *Mattoon (IL) Journal Gazette*, February 17, 1949, 3; "Ship Railway Study Complete," *Reno (NV) Evening Gazette*, April 17, 1949, 8.

58. "Competition Spurs Drive by Free Port," *New York Times*, August 28, 1940, 57; "New Orleans Boasts Second Foreign Trade Zone in U.S.," *Wall Street Journal*, May 2, 147, 1.

59. M. Rolland, *Stationary Dredge*, 4.

60. M. Rolland, *Stationary Dredge*, 6–14.

61. M. Rolland, *Stationary Dredge*, 18.

12. Out of the Ports and into the Hills

1. Gauss, *Made in Mexico*, 188.

2. Rolland, *Efectiva manera*, 1–2.

3. Rolland, *Efectiva manera*, 2–11.

4. R. Alexander, *Sons of the Mexican Revolution*, 102–3.

5. R. Alexander, *Sons of the Mexican Revolution*, 85; B. Smith, "Building a State on the Cheap," 255–70.

6. George's arguments continue to have some influence on economists. See, for examples, Elizabeth Lesly Stevens, "A Tax Policy with San Francisco Roots," *New York Times*, July 30, 2011, http://www.nytimes.com/2011/07/31/us/31bcstevens.html?_r=0; E.S.L, "Why Henry George Had a Point," *The Economist*, April 1, 2015, http://www.economist.com/blogs/freeexchange/2015/04/land-value-tax; and B. Smith, "Building a State on the Cheap," 257–62.

7. Other prominent figures in the Mexican Revolution influenced by Henry George include Marte R. Gómez, Ramón P. de Negri, and Manuel N. Robles.

8. Sydney Gruson, "Protest in Oaxaca on Governor Gains," *New York Times*, March 27, 1952, 15.

9. These critiques of the Mexican Revolution are quoted in Ross, *Is the Mexican Revolution Dead?*, 94.

10. Sydney Gruson, "Another Peaceful Mexican Election Forecast After Nationwide Survey," *New York Times*, July 4, 1952, 6.

11. Ruiz Cortines quoted in Krauze, *Mexico*, 601.

12. Krauze, *Mexico*, 607–8.

13. *Turco*, literally meaning Turkish, was a usually derogatory term for people with Middle Eastern heritage.

14. Jesús Hernández to Adolfo Ruiz Cortines, September 30, 1953, AGN, RPR, Ruiz Cortines, caja 1265, exp. 703.6/14; Luis García Larrañaga to Antonio Carillo Flores, October 22, 1953, AGN, RPR, Ruiz Cortines, caja 1265, exp. 703.6/14.

15. Jesús Hernández to Adolfo Ruiz Cortines, September 30, 1953, AGN, RPR, Ruiz Cortines, caja 1265, exp. 703.6/14.

16. "Ban on U.S. Whiskey Lifted by Mexico," *New York Times*, July 22, 1953, 8.

17. Junta Directiva del Puertos Libres Mexicanos to Ramón Beteta, March 25, 1947, AGN, RPR, Miguel Alemán Valdés, caja 0903, exp. 316/21.

18. In Tehuantepec resistance to development schemes and foreign invasion predates the arrival of the Spanish, but it is not well documented until the second half of the twentieth century. It is likely that similar concerns occurred during Rolland's time in the region. See Call, *No Word for Welcome*.

19. Modesto C. Rolland to Adolfo Ruiz Cortines, February 2, 1955, AGN, RPR, Ruiz Cortines, caja 1265, exp. 703.6/14; Modesto C. Rolland to Adolfo Ruiz Cortines, May 9, 1955, AGN, RPR, Ruiz Cortines, caja 1265, exp. 703.6/14.

20. "Mexico," *La Prensa* (San Antonio TX), March 15, 1957, 6; Enrique Sodi Álvarez, "Istmo de Tehuantepec," *Excélsior* (Mexico City), March 21, 1968, 4.

21. J. Rolland, "Un constructor del México moderno," 84; Jorge M. Rolland, email to J. Justin Castro, September 18, 2016.

22. J. Rolland, "Un constructor del México moderno," 84.

23. J. Rolland, "Modesto C. Rolland: Vida y obra," 31.

24. Beals, *Latin America*, 14.

25. Ruíz, *Mexico*, 157.

26. J. Rolland, "Rancho la Santa Margarita, Córdoba, Veracruz," unpublished document, JMR.

27. J. Rolland, "Rancho la Santa Margarita, Córdoba, Veracruz," unpublished document, JMR.

28. Modesto Rolland, "Vida y hechos del Ing. M. C. Rolland," unpublished document, 1964, JMR.

29. J. Rolland, "Rancho la Santa Margarita, Córdoba, Veracruz," unpublished document, JMR.

30. Jorge M. Rolland, email to J. Justin Castro, October 18, 2016; Deanna Wicks, email to Jorge M. Rolland and J. Justin Castro, October 18, 2016.

Conclusion

1. Sun Yat-sen, the first president of the post–Qing dynasty Republic of China, "was deeply impressed" by Henry George's single tax. Schrecker, *Chinese Revolution*, 171.

2. Naldy Rodríguez, "Estadio emblema de Xalapa cumplirá 90 años," *El Universal Veracruz*, August 17, 2015, http://www.eluniversalveracruz.com.mx/desarrollo-sociedad/2015/estadio-emblema-de-xalapa-cumplira-90-anos-22219.html.

3. Ángel Rafael Martínez Alarcón, "Conversatorio . . . Estadio Jalapeño, sus 90 años," *Tribuna Libre*, August 18, 2015, http://www.tribunalibrenoticias.com/2015/09/conversatorio-estadio-jalapeno-sus-90.html.

4. Eligio Moisés Coronado, "Personajes Célebres Sudcalifornios: Modesto C. Rolland Mejía," Sudcalifornios.com, February 9, 2018, http://sudcalifornios.com/item/personajes-celebres-sudcalifornios-modesto-c-rolland-mejia; Calvillo Velasco, "Prólogo"; J. Rolland, *Modesto C. Rolland*.

5. Lauren Cocking, "Iconic Mexico City Football Stadium to Be Demolished," Culture Trip, August 30, 2016, https://theculturetrip.com/north-america/mexico/articles/iconic-mexico-city-football-stadium-to-be-demolished; "El Estadio Azul será demolido y Cruz Azul se va al Azteca," Univision, August 3, 2016, http://www.univision.com/deportes/futbol/cruz-azul/el-estadio-azul-sera-demolido-y-cruz-azul-se-va-al-azteca.

6. See Garner, "Politics of National Development," 339–56; Garner, *British Lions and Mexican Eagles*.

7. Bulnes, *Los grandes problemas de México*, 286–90.

8. A. Barrios Gómez, "Tehuantepec competirá," *El Heraldo de México*, July 6, 1966, 1; Winberry, "Mexican Landbridge Project," 12; Jorge Áviles Randolph, "Proyectan un ferrocarril de vía doble, electrificado, en el Istmo," *El Universal* (Mexico City), March 20, 1974, 1, 7; "Big Isthmus Project Set to Start Next Year," *Mexico Daily News*, June 26, 2015, http://mexiconewsdaily.com/news/big-isthmus-project-set-to-start-next-year.

9. "Global Poverty Declines Even amid Economic Slowdown," National Public Radio, October 2, 2016, http://www.npr.org/sections/thetwo-way/2016/10/02/496099777/global-poverty-declines-even-amid-economic-slowdown-world-bank-says; "Remarkable Declines in Global Poverty, but Major Challenges Remain," World Bank, April 17, 2013, http://www.worldbank.org/en/news/press-release/2013/04/17/remarkable-declines-in-global-poverty-but-major-challenges-remain.

10. "Oxfam Says Wealth of Richest 1% Equal to Other 99%," BBC News, January 18, 2016, http://www.bbc.com/news/business-35339475; Larry Elliott, "World's Eight Richest People Have Same Wealth as Poorest 50%," *The Guardian* (UK), January 15, 2017, https://www.theguardian.com/global-development/2017/jan/16/worlds-eight-richest-people-have-same-wealth-as-poorest-50?CMP=Share_iOSApp_Other.

11. "Global Climate Change," National Aeronautics and Space Administration, accessed October 31, 2016, http://climate.nasa.gov/.

Bibliography

Archives

ACNSLP	Archivo de la Catedral de Nuestra Señora de La Paz, Baja California Sur
AGEH	Archivo General de la Universidad Autónoma del Estado de Hidalgo, Pachuca
AH	Archivo Histórico
AGEV	Archivo General del Estado de Veracruz, Xalapa
G	Gobernación
PE	Poder Ejecutivo
AGEY	Archivo General del Estado de Yucatán, Mérida
PE	Poder Ejecutivo
G	Gobernación
HP	Hacienda Pública
AGN	Archivo General del Nación de México, Mexico City DF
G	Gobernación
H	Hemeroteca
IP	Instrucción Pública y Bellas Artes
RP	Particulares
RPR	Presidentes
P	Patentes
TSJDF	Tribunal Superior de Justicia del Distrito Federal
AHADB	Archivo Histórico Antigua Dirección El Boleo, Santa Rosalía, Baja California Sur
AHPLM	Archivo Histórico del Pablo L. Martínez, La Paz, Baja California Sur
G	Guerra
IP	Instrucción Pública y Bellas Artes
J	Justicia
REG	Regímenes
RP	Porfiriato
RR	Revolución

AHPM	Archivo Histórico del Palacio de Minería, Mexico City DF
ASRE	Archivo Histórico Genaro Estrada de la Secretaría de Relaciones Exteriores, Mexico City DF
CEHM	Centro de Estudios de Historia de México Carso Fundación Carlos Slim, Mexico City DF
FAPECFT	Fideicomiso Archivos Plutarco Elías Calles y Fernando Torreblanca, Mexico City DF
APEC	Archivo Plutarco Elías Calles
FJA	Fondo Joaquín Amaro
FP	Fondo Presidentes
FJM	Archivo Francisco J. Múgica, Centro de Estudios de la Revolución Mexicana "Lázaro Cárdenas," Jiquilpán de Juárez, Michoacán
FFJM	Fondo Francisco J. Múgica
JMR	Jorge M. Rolland C. Private Collection, Querétaro, Querétaro
NARA	U.S. National Archives and Records Administration, College Park, Maryland
RG 59	Record Group 59, Department of State Central Files
RG 84	Record Group 84, Records of the Foreign Service Posts of the Department of State
NLBL	Nettie Lee Benson Library, Latin American Collection, University of Texas at Austin
MRNC	Mexican Revolution Newspaper Collection
UP	Urquidi Papers

Published Works

Aboites Aguilar, Luis. *El agua de la nación: Una historia política de México (1888–1946)*. Mexico City: Centro de Investigaciones y Estudios Superiores en Antropología Social, 1998.

———. *Excepciones y privilegios: Modernización tributaria y centralización en México, 1922–1972*. Mexico City: El Colegio de México, Centro de estudios históricos, 2003.

———, ed. *Pablo Bistráin, ingeniero mexicano*. Mexico City: Centro de Investigaciones y Estudio Superiores en Antropología Sociales, 1997.

Aboites Aguilar, Luis, and Luis Jáuregui, eds. *Penuria sin fin: Historia de los impuestos en México, siglos XVIII–XX*. Mexico City: Instituto Mora, 2005.

Adas, Michael. *Machines as the Measure of Men: Science, Technology, and Ideologies of Western Dominance*. Ithaca NY: Cornell University Press, 1989.

Agostoni, Claudia. *Monuments of Progress: Modernization and Public Health in Mexico City, 1876–1910*. Calgary AB: University of Calgary Press, 2003.

Albarrán, Elena Jackson. *Seen and Heard in Mexico: Children and Revolutionary Cultural Nationalism*. Lincoln: University of Nebraska Press, 2015.

Alegre, Robert F. *Railroad Radicals in Cold War Mexico: Gender, Class, and Memory.* Lincoln: University of Nebraska Press, 2013.

Alemán Valdés, Miguel. *Remembranzas y testimonios.* Mexico City: Editorial Grijalbo, 1986.

Alexander, Anna Rose. *City on Fire: Technology, Social Change, and the Hazards of Progress in Mexico City, 1860–1910.* Pittsburgh: University of Pittsburgh Press, 2016.

———. "Safety by Design: Engineers and Entrepreneurs Invent Fire Safety in Mexico City, 1860–1910." *Urban History* 40, no. 3 (2014): 435–55.

Alexander, Ryan. *Sons of the Mexican Revolution: Miguel Alemán and His Generation.* Albuquerque: University of New Mexico Press, 2016.

Alvarado, Salvador. *Carta al pueblo Yucatán: Mi sueño.* 1916. Mexico City: Maldonado Editores, 1988.

———. *La reconstrucción de México: Un mensaje a los pueblos de América.* 3 vols. Mexico City: J. Ballesca, 1919.

———. *Salvador Alvarado: Pensamiento revolucionario.* 1916. Mérida YUC: Instituto de Seguridad Social de los Trabajadores del Estado de Yucatán, 1980.

Alzanti, Servando A. *Historia de la mexicanización de los Ferrocarriles Nacionales de México.* Mexico City: Beatriz de Silva, 1946.

Anderson, Mark Cronlund. *Pancho Villa's Revolution by Headlines.* Norman: University of Oklahoma Press, 2001.

Austin, Mary Hunter. "What the Mexican Conference Really Means." Press release, Latin-American News Association, 1916.

Ayala Espino, José. *Estado y desarrollo: La formación de la economía mixta mexicana (1920–1982).* Mexico City: Fondo Cultura Económica, 1988.

Babb, Sarah L. *Managing Mexico: Economists from Nationalism to Neoliberalism.* Princeton: Princeton University Press, 2004.

Bailey, David C. *¡Viva Cristo Rey! The Cristero Rebellion and the Church-State Conflict in Mexico.* Austin: University of Texas Press, 1974.

Balcázar Antonio, Elías. *Tabasco a dos tiempos, 1940–1960.* Villahermosa: Universidad Juárez Autónoma de Tabasco, 2011.

Banister, Jeffrey M., and Stacie G. Widdifield. "The Debut of 'Modern Water' in Early 20th Century Mexico City: The Xochimilco Potable Waterworks." *Journal of Historical Geography* 46 (October 2014): 36–52.

Baptista González, David, and Juan José Saldaña. "La participación política y reivindicación gremial del Centro de Ingenieros de México ante la construcción del Estado mexicano en los años veinte." In *Memorias del Primer Coloquio Latinoamericano de Historia y Estudios Sociales sobre la Ciencia y la Tecnología*, edited by Federico Lazarín Miranda, 1221–30. Mexico City: Sociedad Mexicana de Historia de la Ciencia y de la Tecnología, 2007.

Barkin, Daniel. "Mexico's Albatross: The U.S. Economy." In *Modern Mexico: State, Economy, and Social Conflict*, edited by Nora Hamilton and Timothy F. Harding, 106–27. Newbury Park CA: Sage Publications, 1986.

Bazant, Mílada. "La enseñanza y la práctica de la ingeniería durante el Porfiriato." *Historia Mexicana* 33, no. 1 (1984): 254–97.

Beals, Carleton. *Latin America: World in Revolution*. New York: Abelard-Schuman, 1963.

Beatty, Edward. *Technology and the Search for Progress in Modern Mexico*. Oakland: University of California Press, 2015.

Benjamin, Thomas. "Rebuilding the Nation." In *The Oxford History of Mexico*, edited by Michael C. Meyer and William H. Beezley, 467–502. Oxford: Oxford University Press, 2000.

Berman, Marshall. *All That Is Solid Melts into Air: The Experience of Modernity*. New York: Penguin Books, 1982.

Bernstein, Daniel Eli. *Next to Godliness: Confronting Dirt and Despair in Progressive Era New York City*. Urbana: University of Illinois Press, 2006.

Bernstein, Marvin D. *The Mexican Mining Industry, 1890–1950: A Study of Politics, Economics, and Technology*. Albany: State University of New York Press, 1964.

Bijker, Wiebe E., Thomas P. Hughes, and Trevor Pinch, eds. *The Social Construction of Technological Systems: New Directions in the Sociology and History of Technology*. Cambridge MA: MIT Press, 2012.

Bond, Paul Stanley. *The Engineer in War, with Special Reference to the Training of the Engineer to Meet the Military Obligations of Citizenship*. New York: McGraw-Hill, 1916.

Boyd, Consuelo. "Twenty Years to Nogales: The Building of the Guaymas-Nogales Railroad, *Journal of Arizona History* 22, no. 3 (1981): 295–324.

Boyer, Christopher R. "The Cycles of Mexican Environmental History." In *A Land between Waters: Environmental Histories of Modern Mexico*, edited by Christopher R. Boyer, 1–21. Tucson: University of Arizona Press, 2012.

———. "Old Loves, New Loyalties: Agrarismo in Michoacán, 1920–1928." *Hispanic American Historical Review* 78, no. 3 (1998): 419–55.

Brading, David. *Caudillo and Peasant in the Mexican Revolution*. Cambridge: Cambridge University Press, 1980.

———. *Prophecy and Myth in Mexican History*. Cambridge: Cambridge University Press, 1984.

Britton, John A. *Revolution and Ideology: Images of the Mexican Revolution in the United States*. Lexington: University Press of Kentucky, 1995.

Brooking, Tom. *The History of New Zealand*. Santa Barbara CA: Greenwood Press, 2004.

Brooking, Tom, and Eric Pawson. *Seeds of Empire: The Environmental Transformation of New Zealand*. London: I. B. Tauris, 2010.

Brunk, Samuel. ¡Emiliano Zapata! *Revolution and Betrayal in Mexico*. Albuquerque: University of New Mexico Press, 1995.

Buchenau, Jürgen. *Plutarco Elías Calles and the Mexican Revolution*. Lanham MD: Rowman and Littlefield, 2006.

Buffington, Robert M., and William E. French. "The Culture of Modernity." In *The Oxford History of Mexico*, edited by Michael C. Meyer and William H. Beezley, 397–432. New York: Oxford University Press, 2000.

Bulnes, Francisco. *Los grandes problemas de México*. Mexico City: Ediciones de "El Universal," 1927.

Bunker, Steven. *Creating Mexican Consumer Culture in the Age of Porfirio Díaz*. Albuquerque: University of New Mexico Press, 2012.

Burns, Edward McNall. *David Star Jordan: Prophet of Freedom*. Stanford: Stanford University Press, 1953.

Call, Wendy. *No Word for Welcome: The Mexican Village Faces the Global Economy*. Lincoln: University of Nebraska Press, 2011.

Calles, Plutarco Elías. *Pensamiento político y social: Antología*. Edited by Carlos Macías Richard. Mexico City: Fondo de Cultura Económica, 1988.

Calvillo Velasco, Max. "Indicios para descifrar la trayectoria política de Esteban Cantú." *Historia Mexicana* 59, no. 3 (2010): 981–1039.

———. "Prólogo." In *Informe sobre el Distrito Norte de Baja California*, by Modesto C. Rolland. Mexico City: Secretaria de Educación Pública y la Universidad Autónoma de Baja California, 1993.

Cámara Nacional del Cemento. *Medio siglo de cemento en México*. Mexico City, 1963.

Cámara Zavala, Gonzalo. *Reseña histórica de la industria henequenera de Yucatán*. Mérida YUC: Imp. Oriente, 1936.

Candiani, Vera S. *Dreaming of Dry Land: Environmental Transformation in Colonial Mexico City*. Stanford: Stanford University Press, 2014.

Cárdenas, Lázaro. "Al Gral. Lázaro Cárdenas, al abrir el Congreso sus sesiones ordinarias, el 1º de septiembre de 1935." *Los presidentes de México ante la Nación: Informes, manifiestos y documentos de 1821 a 1966*. Edited by the XLVI Legislatura de la Cámara de Diputados. Mexico City: Imprenta de la Cámara de Diputados, 1966.

Cariño, Micheline, and Mario Monteforte. "History of Pearling in La Paz Bay, South Baja California." *Gems and Gemology* 30, no. 2 (1995): 88–105.

Carr, Barry. *Marxism and Communism in Twentieth-Century Mexico*. Lincoln: University of Nebraska Press, 1992.

Carranza, Luis E. *Architecture as Revolution: Episodes in the History of Modern Mexico*. Austin: University of Texas Press, 2010.

Carranza, Venustiano. *Report by Venustiano Carranza (First Chief of the Constitutionalist Army) in the City of Querétaro, State of Querétaro, Méx., Friday, December 1st, 1916*. New York: Latin-American News Association, 1916.

Castañeda Crisolis, Edgar, and Juan José Saldaña. "El *Boletín del Petróleo*: Una enciclopedia tecnológica para la industria petrolera mexicana (1916–1933)." *Quipu* 15, no. 1 (2013): 85–100.

Castillo Peraza, Carlos, ed. *Manuel Gómez Morín, constructor de instituciones*. Mexico City: Fondo de Cultura Económica, 1994.

Castro, J. Justin. "Modesto C. Rolland and the Development of Baja California." *Journal of the Southwest* 58, no. 2 (2016): 261–92.

——— . *Radio in Revolution: Wireless Technology and State Power in Mexico, 1897–1938*. Lincoln: University of Nebraska Press, 2016.

——— . "Radiotelegraphy to Broadcasting: Wireless Communications in Porfirian and Revolutionary Mexico, 1899–1924." *Mexican Studies/Estudios Mexicanos* 29, no. 2 (2013): 335–65.

——— . "Sounding the Mexican Nation: Intellectuals, State Building, and the Culture of Early Radio Broadcasting." *Latin Americanist* 58, no. 3 (2014): 3–30.

Castro, Pedro. *Álvaro Obregón: Fuego y cenizas de la Revolución Mexicana*. Mexico City: Ediciones Era, 2010.

Castro Martínez, Pedro Fernando. *Adolfo de la Huerta: La integridad como arma de la revolución*. Mexico City: Siglo Veintiuno Editores, 1998.

Chacón, Ramón D. "Salvador Alvarado and Agrarian Reform in Yucatán, 1915–1918: Federal Obstruction of Regional Social Change." In *Land, Labor, and Capital in Modern Yucatán: Essays in Regional History and Political Economy*, edited by Jeffrey T. Brannon and Gilbert M. Joseph, 179–96. Tuscaloosa: University of Alabama Press, 1991.

Coatsworth, John H. *Growth against Development: The Economic Impact of Railroads in Porfirian Mexico*. DeKalb: Northern Illinois University Press, 1981.

Cockcroft, James D. *Intellectual Precursors of the Mexican Revolution, 1900–1913*. Austin: University of Texas Press, 1968.

——— . *Mexico's Revolution: Then and Now*. New York: New York University Press, 2010.

Coleman, Peter J. *Progressivism and the World of Reform: New Zealand and the Origins of the American Welfare State*. Lawrence: University Press of Kansas, 1987.

Connolly, Priscilla. *El contratista de don Porfirio: Obras públicas, deuda y desarrollo desigual*. Mexico City: Colegio de Michoacán, UAMA, Fondo de Cultura Económica, 1997.

Cook, Blanche W. *Crystal Eastman on Women and Revolution*. Oxford: Oxford University Press, 1978.

Craib, Raymond B. *Cartographic Mexico: A History of State Fixations and Fugitive Landscapes*. Durham: Duke University Press, 2004.

Cruz, Salvador. *Vida y obra de Pastor Rouaix*. Mexico City: Instituto Nacional de Antropología e Historia, 1980.

Cumberland, Charles Curtis. "Genesis of Mexican Agrarian Reform." *The Historian* 14, no. 2 (1952): 209–32.

———. *Mexican Revolution: Genesis under Madero*. Austin: University of Texas Press, 1952.

Daniels, Doris Groshen. *Always a Sister: The Feminism of Lillian D. Wald*. New York: Feminist Press at CUNY, 1995.

Dawson, Alexander S. *Indian and Nation in Revolutionary Mexico*. Tucson: University of Arizona Press, 2004.

Deeds, Susan. "José María Maytorena and the Mexican Revolution in Sonora (Part II)." *Arizona and the West* 18, no. 2 (1976): 125–48.

de Gortari Rabiela, Rebeca. "Educación y conciencia nacional: Los ingenieros después de la revolución mexicana." *Revista Mexicana de Sociología* 49, no. 3 (1987): 123–41.

de Ibarrola, José Rámon. *Apuntes sobre el desarrollo de la ingeniería en México y la educación del ingeniero*. Mexico City: Tipografía de la viuda de F. Díaz de León, 1911.

de la Paz Ramos Lara, María, and Rigoberto Rodríguez Benítez, eds. *Formación de ingenieros en el México del siglo XIX*. Mexico City: Universidad Nacional Autónoma de México, 2007.

Delpar, Helen. *The Enormous Vogue of Things Mexican: Cultural Relations between the United States and Mexico, 1920–1935*. 1992. Tuscaloosa: University of Alabama Press, 2015. eBook.

di Fornaro, Carlo. With chapters by I. C. Enriquez, Charles Ferguson, and M. C. Rolland. *Carranza and Mexico*. New York: Mitchell Kennerley, 1915.

Domínguez Martínez, Raúl. *La ingeniería civil en México, 1900–1940: Análisis histórico de los factores de su desarrollo*. Mexico City: Universidad Nacional Autónoma de México, 2013.

Domínguez Tapa, Carlos. *Forjadores de Baja California*. Mexico City: Editorial Aristos, 1980.

Dorsey, Florence. *Road to the Sea: The Story of James B. Eads and the Mississippi River*. New York: Rinehart and Company, 1947.

Dulles, John W. F. *Yesterday in Mexico: A Chronicle of the Revolution, 1919–1936*. Austin: University of Texas Press, 1961.

Eineigel, Susanne. "Revolutionary Promises Encounter Urban Realities for Mexico City's Middle Class, 1915–1928." In *The Making of the Middle Class: Toward a Transnational History*, edited by A. Ricardo López and Barbara Weinstein, 253–66. Durham: Duke University Press, 2012.

Eisenhower, John S. D. *Intervention! The United States and the Mexican Revolution, 1913–1917*. New York: Norton, 1993.

Eiss, Paul K. *In the Name of El Pueblo: Place, Community, and the Politics of History in Yucatán*. Durham: Duke University Press, 2010.

Ervin, Michael A. "The Formation of the Revolutionary Middle Class during the Mexican Revolution." In *The Making of the Middle Class: Toward a Transnational History*, edited by A. Ricardo López and Barbara Weinstein, 196–222. Durham: Duke University Press, 2012.

Evans, Sterling. *Bound in Twine: The History and Ecology of the Henequen-Wheat Complex for Mexico and the Canadian Plains, 1880–1950*. College Station: Texas A&M University Press, 2007.

———. "King Henequen: Order, Progress, and Ecological Change in Yucatán." In *A Land between Waters: Environmental Histories of Modern Mexico*, edited by Christopher R. Boyer. Tucson: University of Arizona Press, 2012.

Fallaw, Ben. *Religion and State Formation in Postrevolutionary Mexico*. Durham: Duke University Press, 2013.

Ferguson, Niall. *The House of Rothschild*. Volume 1, *Money's Prophets: 1798–1848*. New York: Penguin Books, 1999.

Fernando, Benítez. *Lázaro Cárdenas y la Revolución mexicana*. Volume 3, *El cardenismo*. Mexico City: Fondo de Cultura Económica, 1998.

Ferrer Mendiolea, Gabriel. *Historia del Congreso Constituyente de 1916–1917*. Mexico City: Talleres Gráficos de la Nación, 1957.

Fischer, David Hackett. *Fairness and Freedom: A History of Two Open Societies, New Zealand and the United States*. New York: Oxford University Press, 2012.

Flint, Anthony. *Modern Man: The Life of Le Corbusier, Architect of Tomorrow*. Boston: New Harvest, 2014.

Flores, Ruben. *Backroads Pragmatists: Mexico's Melting Pot and Civil Rights in the United States*. Philadelphia: University of Pennsylvania Press, 2014.

Flores, Trinidad W. *Contraespionaje político y sucesión presidencial: Correspondencia de Trinidad W. Flores sobre la primera campaña electoral de Álvaro Obregón*. Mexico City: Universidad Autónoma de México, 1985.

Flores Ayala, Hubonor. "Adalberto Tejada, biografía de un agrarista radical." In *Veracruzanos en la independencia y la revolución*, edited by Abel Juárez Martínez. Xalapa: Veracruz Gobierno del Estado, 2010.

Fowler, Will. *Santa Anna of Mexico*. Lincoln: University of Nebraska Press, 2009.

Free Ports Executive Board. *Mexican Free Ports*. Mexico City: Revista de Hacienda, 1924.

Gallo, Rubén. *Mexican Modernity: The Avant-Garde and the Technological Revolution*. Cambridge MA: MIT Press, 2005.

Garciadiego Dantán, Javier. *Rudos contra científicos: La Universidad Nacional durante la Revolución Mexicana*. Mexico City: Colegio de México and Universidad Nacional Autónoma de México, 1996.

Garner, Paul. *British Lions and Mexican Eagles: Business, Politics, and Empire in the Career of Weetman Pearson in Mexico, 1889-1919*. Stanford: Stanford University Press, 2011.

———. "The Politics of National Development in Late Porfirian Mexico: The Reconstruction of the Tehuantepec National Railway, 1896-1907." *Bulletin of Latin American Research* 14, no. 3 (1995): 339-56.

———. *Porfirio Díaz*. New York: Longman, 2001.

Gauss, Susan. *Made in Mexico: Regions, Nation, and the State in the Rise of Mexican Industrialism, 1920s-1940s*. University Park: Pennsylvania State University Press, 2010.

George, Henry. *Our Land and Land Policy*. 1901. Lansing: Michigan State University, 1991.

———. *Progress and Poverty: An Inquiry in the Cause of Industrial Depressions and of Increase in Want with Increase of Wealth*. 1879. New York: Robert Schakenbach Foundation, 1997.

Gereffi, Gary, and Donald L. Wyman. *Manufacturing Miracles: Paths of Industrialization in Latin America and East Asia*. Princeton: Princeton University Press, 1990.

Giedion, Siegfried. *Walter Gropius*. New York: Dover, 1992

Gilderhus, Mark T. *The Second Century: U.S.-Latin American Relations since 1889*. New York: Rowman and Littlefield, 1999.

Gillette, Howard, Jr. *Civitas by Design: Building Better Communities, from Garden Cities to the New Urbanism*. Philadelphia: University of Pennsylvania Press, 2010.

Gilly, Adolfo. *The Mexican Revolution*. Translated by Patrick Camiller. New York: New Press, 2005.

Gonzales, Michael J. *The Mexican Revolution, 1910-1940*. Albuquerque: University of New Mexico Press, 2002.

Grubb, Kenneth. "The Political and Religious Situation in Mexico." *International Affairs* 14, no. 5 (1935): 674-94.

Grunstein Dickter, Arturo. "In the Shadow of Oil: Francisco J. Múgica vs. Telephone Transnational Corporation in Cardenista Mexico." *Mexican Studies/Estudios Mexicanos* 21, no. 1 (2005): 1-32.

Guzmán, Martín Luis. *El águila y la serpiente*. 1928. Mexico City: Casiopera, 2000.

Haber, Stephen. *Industry and Underdevelopment: The Industrialization of Mexico, 1900-1940*. Stanford: Stanford University Press, 1989.

Haber, Stephen, Armando Razo, and Noel Maurer. *The Politics of Property Rights: Political Instability, Credible Commitments, and Economic Growth in Mexico, 1876-1929*. Cambridge: Cambridge University Press, 2004.

Hale, Charles. *Emilio Rabasa and the Survival of Porfirian Liberalism: The Man, His Career, and His Ideas, 1856-1930*. Stanford: Stanford University Press, 2008.

———. "Frank Tannenbaum and the Mexican Revolution." *Hispanic American Historical Review* 75, no. 2 (1995): 215–46.

Haley, P. Edward. *Revolution and Intervention: The Diplomacy of Taft and Wilson with Mexico, 1910–1917*. Cambridge MA: MIT Press, 1970.

Hall, Linda B. "Alvaro Obregón and the Politics of Mexican Land Reform, 1920–1924." *Hispanic American Historical Review* 60, no. 2 (1980): 213–238.

———. *Álvaro Obregón: Power and Revolution in Mexico, 1911–1920*. College Station: Texas A&M University Press, 1981.

Hall, Peter. *Cities of Tomorrow: An Intellectual History of Urban Planning and Design in the Twentieth Century*. Cambridge MA: Basic Blackwell, 1988.

Hamon, James L., and Stephen R. Niblo. *Precursores de la revolución agraria en México: Las obras de Wistano Luis Orozco y Andrés Molina Enríquez*. Mexico City: Secretaría de Educación Pública, 1975.

Harris, Charles H., III, and Luis R. Sadler. "The 'Underside' of the Mexican Revolution: El Paso, 1912." *The Americas* 39, no. 1 (1982): 69–83.

Hart, John Mason. *Anarchism and the Mexican Working Class, 1860–1931*. Austin: University of Texas Press, 1978.

———. *Empire and Revolution: The Americans in Mexico since the Civil War*. Berkeley: University of California Press, 2002.

———. *Revolutionary Mexico: The Coming and Process of the Mexican Revolution*. Berkeley: University of California Press, 1987.

Hart, Paul. *Bitter Harvest: The Social Transformation of Morelos, Mexico, and the Origins of the Zapatista Revolution, 1840–1910*. Albuquerque: University of New Mexico Press, 2007.

Hayes, Joy. *Radio Nation: Communication, Popular Culture, and Nationalism in Mexico, 1920–1950*. Tucson: University of Arizona Press, 2000.

Headrick, Daniel R. *The Invisible Weapon: Telecommunications and International Politics, 1851–1945*. New York: Oxford University Press, 1991.

———. *The Tentacles of Progress: Technology Transfer in the Age of Imperialism, 1850–1940*. New York: Oxford University Press, 1988.

———. *The Tools of Empire: Technology and European Imperialism in the Nineteenth Century*. New York: Oxford University Press, 1981.

Henderson, Timothy J. *The Worm in the Wheat: Rosalie Evans and Agrarian Struggle in the Puebla-Tlaxcala Valley of Mexico, 1906–1927*. Durham: Duke University Press, 1998.

Hernández Palacios, Aureliano. *Xalapa de mis recuerdos*. Xalapa: Gobierno del Estado de Veracruz-Llano, 1986.

Hershfield, Joanne. *Imagining la Chica Moderna: Women, Nation, and Visual Culture in Mexico, 1917–1936*. Durham: Duke University Press, 2008.

Hill, Jonathan, Jr. "Circuits of State: Water, Electricity, and Power in Chihuahua, 1905–1936." *Radical History Review* 127, no. 1 (2017): 13–38.

Horn, James J. "U.S. Diplomacy and the 'Specter of Bolshevism' in Mexico (1924–1927)." *The Americas* 32, no. 1 (1975): 31–45.

Howard, Ebenezer. *Garden Cities of To-morrow*. 2nd ed. London: Swan Sonnenschein, 1902.

Howe, Jerome W. *Campaigning in Mexico, 1916: Adventures of a Young Officer in General Pershing's Punitive Expedition*. Tucson: Arizona Historical Society, 1968.

Ibarra Rivera, Gilberto. *Historia de la educación en Baja California Sur*. Volume 1, *Desde la colonia hasta el siglo XIX*. La Paz: Benemérita Escuela Normal Urbana, 1993.

Jacoby, Karl. *The Strange Career of William Ellis: The Texas Slave Who Became a Mexican Millionaire*. New York: Norton, 2016.

Jordan, David Starr. *The Days of a Man: Being Memories of a Naturalist, Teacher, and Minor Prophet of Democracy*. Volume 2, *1900–1912*. New York: World Book Company, 1922.

———. "Patriotism, Nationalism and Peace." *Advocate for Peace* 78, no. 2 (1916): 43–45.

Joseph, Gilbert M. *Revolution from Without: Yucatán, Mexico, and the United States, 1880–1924*. Durham: Duke University Press, 1988.

Joseph, Gilbert M., and Jürgen Buchenau. *Mexico's Once and Future Revolution: Social Upheaval and the Challenge of Rule since the Late Nineteenth Century*. Durham: Duke University Press, 2013.

Josephson, Paul R. *Industrialized Nature: Brute Force Technology and the Transformation of the Natural World*. Washington DC: Island Press, 2002.

Kanellos, Nicolás. "Cronistas and Satire in Early Twentieth Century Hispanic Newspapers." *MELUS* 23, no. 1 (1998): 3–25.

Katz, Friedrich. *The Life and Times of Pancho Villa*. Stanford: Stanford University Press, 1998.

———. *The Secret War in Mexico: Europe, the United States, and the Mexican Revolution*. Chicago: University of Chicago Press, 1981.

Kellogg, Paul U. "A New Era of Friendship for North America." *The Survey* 36, no. 13 (July 15, 1916): 415–17.

Kerber, Víctor. "El supuesto lo complot nipo-mexicano contra Estados Unidos durante la Revolución." *Estudios de Asia y Africa* 27, no. 1 (1992): 28–50.

Klich, Lynda. "Estridentópolis: Achieving a Post-Revolutionary Utopia in Jalapa." Special issue, "Mexico," *Journal of Decorative and Propaganda Arts* 26 (April 2010): 102–27.

Kloppenberg, James T. *Uncertain Victory: Social Democracy and Progressivism in European and American Thought, 1870–1920*. New York: Oxford University Press, 1988.

———. *The Virtues of Liberalism*. New York: Oxford University Press, 2000.

Knight, Alan. "Land and Society in Revolutionary Mexico: The Destruction of the Great Haciendas." *Mexican Studies/Estudios Mexicanos* 7, no. 1 (1991): 73–104.

———. *The Mexican Revolution*. 2 vols. Lincoln: University of Nebraska Press, 1986.

Krauze, Enrique. *Mexico, Biography of Power: A History of Modern Mexico, 1810–1996*. Translated by Hank Heifetz. New York: HarperCollins, 1997.

———. *Venustiano Carranza, puente entre siglos*. Mexico City: Fondo de Cultura Económica, 1987.

Kroeber, Clifton B. *Man, Land, and Water: Mexico's Farmlands Irrigation Policies, 1885–1911*. Berkeley: University of California Press, 1983.

Kuntz Ficker, Sandra, ed. *Historia mínima de la economía mexicana, 1519–2010*. Mexico City: Colegio de México, 2012.

La Botz, Dan. "American 'Slackers' in the Mexican Revolution: International Proletarian Politics in the Midst of a National Revolution." *The Americas* 62, no. 4 (2006): 563–90.

Lara G., Carlos. *Manuel Gómez Morín, un gestor cultural en la etapa constructiva de la Revolución mexicana*. Mexico City: Miguel Ángel Porrúa, 2011.

Laurent, John, ed. *Henry George's Legacy in Economic Thought*. Cheltenham UK: Edward Elgar, 2005.

Levenstein, Harvey. "Samuel Gompers and the Mexican Labor Movement." *Wisconsin Magazine of History* 51, no. 2 (1967–68): 155–63.

Lewis, Stephen. *Ambivalent Revolution: Forging State and Nation in Chiapas, 1910–1945*. Albuquerque: University of New Mexico Press, 2005.

———. "The Nation, Education, and the 'Indian Problem' in Mexico, 1920–1940." In *The Eagle and the Virgin: Nation and Cultural Revolution in Mexico, 1920–1940*, edited by Mary Kay Vaughan and Stephen E. Lewis, 176–95. Durham: Duke University Press, 2006.

Loza, Mireya. *Defiant Braceros: How Migrant Workers Fought for Racial, Sexual, and Political Freedom*. Chapel Hill: University of North Carolina Press, 2016.

Madhaven, Guru. *Applied Minds: How Engineers Think*. New York: Norton, 2015.

Maples Arce, Manuel. *Soberana juventud*. Madrid: Editorial Plenitud, 1967.

Marroquín y Rivera, Manuel. *Memoria descriptiva de las obras de provisión de aguas potables para la Ciudad de México*. Mexico City: Müller Hnos., 1914.

Martínez, Pablo L. *Guía familiar de la Baja California, 1700–1900*. Mexico City: Ediciones Baja California, 1965.

Martínez Verdugo, Arnoldo. *Historia del comunismo en México*. Mexico City: Colección Enlance, 1983.

Marván Laborde, Ignacio. *La Revolución Mexicana, 1908–1932*. Mexico City: Fondo de Cultura Económica, 2010.

Marx, Leo. *The Machine in the Garden: Technology and the Pastoral Ideal in America*. New York: Oxford University Press, 2000.

Matthews, Michael. *The Civilizing Machine: A Cultural History of Mexican Railroads, 1876–1910*. Lincoln: University of Nebraska Press, 2014.

Matute, Álvaro. *Historia de la Revolución Mexicana, 1917–1924: La carrera del caudillo*. Mexico City: Colegio de México, 1980.

Maurer, Noel. *The Power and the Money: The Mexican Financial System, 1876–1932*. Stanford: Stanford University Press, 2002.

Maxwell, William H. "Attitude of Parents toward Education." *Journal of Education* 77, no. 14 (April 3, 1913): 372–73.

———. *A Quarter Century of Public School Development*. New York: American Book Company, 1912.

McLuhan, Marshall. *Understanding Media: The Extensions of Man*. New York: McGraw-Hill, 1964.

Medina Ávila, Virginia, and Gilberto Vargas Arana. *Nuestra es la voz, de todos la palabra: Historia de la radiodifusión mexicana, 1921–2010*. Mexico City: Universidad Nacional Autónoma de México-FES, Acatlán, 2011.

Mein-Smith, Philippa. *A Concise History of New Zealand*. 2nd ed. New York: Cambridge University Press, 2012.

Mejía Barquera, Fernando. *La industria de la radio y televisión y la política del estado mexicano (1920–1960)*. Mexico City: Fundación Manuel Buendía, 1989.

Mendoza, María del Rosario Juan. "Heriberto Jara Corona, memorias de sus batallas por instaurar une legislación acorde con las necesidades de los trabajadores." In *Veracruzanos en la independencia y la revolución*, edited by Abel Juárez Martínez. Xalapa: Veracruz Gobierno del Estado, 2010.

Mesa Andraca, Manuel. *Con Salvador Alvarado en Yucatán*. Mexico City: G. Murillo, 1978.

Mexican-American Peace Committee. "The Mexican-American League." Press release, Latin-American News Association, 1916.

Mexican Bureau of Information. *"Red Papers" of Mexico: An Exposé of the Great Científico Conspiracy to Eliminate Don Venustiano Carranza: Documents Relating to the Imbroglio between Carranza and Villa*. New York: Mexican Bureau of Information, 1914.

Meyer, Jean. *La Cristiada*. Volume 1, *La guerra de los cristeros*. Mexico City: Siglo XXI, 2001.

Meyer, Lorenzo. *México y los Estados Unidos en el conflicto petróleo, 1917–1942*. Mexico City: Colegio de México, 1972.

Meyer, Michael C. *Huerta*. Lincoln: University of Nebraska Press, 1972.

Meyers, Allan. *Outside the Hacienda Walls: The Archaeology of Plantation Peonage in Nineteenth-Century Yucatán*. Tucson: University of Arizona Press, 2012.

Moore, H. F. "Engineering Culture." *Science*, n.s. 73, no. 1881 (January 16, 1931): 51–54.

Moreno-Brid, Juan Carlos, and Jaime Ros. *Development and Growth in the Mexican Economy: A Historical Perspective*. Oxford: Oxford University Press, 2009.

Niblo, Stephen R. *Mexico in the 1940s: Modernity, Politics, and Corruption.* Wilmington DE: SR Books, 1999.

———. *War, Diplomacy, and Development: The United States and Mexico, 1938-1954.* Wilmington DE: SR Books, 1995.

Niemeyer, E. V., Jr. *Revolution at Querétaro: The Mexican Constitutional Convention of 1916-1917.* Austin: University of Texas Press, 1974.

Nisbet, Robert. *History of the Idea of Progress.* New York: Basic Books, 1980.

Novo, Salvador. *La vida en México en el período presidencial de Lázaro Cárdenas.* Mexico City: Consejo Nacional para la Cultura y las Artes, 1994.

Noyola, Leopoldo. *La raza de la herba: Historia de telégrafos Morse en México.* 2nd ed. Puebla: Benemérita Universidad Autónoma de Puebla, 2004.

Núñez Tapia, Francisco Alberto. "Aspectos del turismo en el Distrito Norte de Baja California, 1920-1929." *Meyibó* 3, no. 6 (2012): 59-60.

Ornelas Herrera, Roberto. "Radio y cotidianidad en México (1900-1930)." In *Historia de la vida cotidiana en México.* Tomo V, volume 1, *Siglo XX, Campo y ciudad*, edited by Aurelio de los Reyes, 127-69. Mexico City: Fondo de Cultura Económica, 2006.

Osorio Marbán, Miguel. *Carranza: Soberanía y petróleo.* Mexico City: Partido Revolucionario Institucional, 1994.

Page, Max. *The Creative Destruction of Manhattan, 1900-1940.* Chicago: University of Chicago, 1999.

Pani, Alberto J. *Mi contribución al nuevo régimen (1910-1933).* Mexico City: Editorial Cultura, 1936.

Paoli, Francisco José. *Yucatán y los orígenes del nuevo estado mexicano.* Mérida YUC: Universidad Autónoma de México, 2001.

Parra, Max. *Writing Pancho Villa's Revolution.* Austin: University of Texas Press, 2005.

Paxman, Andrew. *Jenkins of Mexico: How a Southern Farm Boy Became a Mexican Magnate.* New York: Oxford University Press, 2017.

Paz, María Emilia. *Strategy, Security, and Spies: Mexico and the U.S. as Allies in World War II.* University Park: Pennsylvania State University Press, 1997.

Peniche Rivero, Piedad. "Recordando a Elvia Carrillo Puerto: Efemérides del triunfo de la lucha por el sufragio femenino." Accessed December 12, 2016. http://www.archivogeneral.yucatan.gob.mx/Efemerides/ElviaCarrillo/ElviaCarrilloPuerto.htm.

Pérez, Luis A., Jr. *Cuba under the Platt Amendment, 1902-1934.* Pittsburgh: University of Pittsburgh Press, 1991.

Plasencia de la Parra, Enrique. *Personajes y escenarios de la rebelión delahuertista, 1923-1924.* Mexico City: Miguel Ángel Porrua, 1998.

Pollock, Horatio M., and William S. Morgan. *Modern Cities.* New York: Funk & Wagnalls, 1913.

Post, Louis Freeland. *The Prophet of San Francisco.* New York: Vanguard Press, 1930.

Preciado Llamas, Juan. *En la periferia del régimen: Baja California Sur durante la administración porfiriana.* La Paz: Universidad Autónoma de Baja California Sur, 2005.

Quirk, Robert E. "Liberales y radicales en la Revolución Mexicana." *Historia Mexicana* 2, no. 4 (1953): 503–28.

Raat, W. Dirk. *Revoltosos: Mexico's Rebels in the United States, 1903–1923.* College Station: Texas A&M University Press, 1981.

Rae, Jon, and Rudi Volti. *The Engineer in History.* New York: Peter Lang, 1993.

Rankin, Monica. *¡México, la Patria! Propaganda and Production during World War II.* Lincoln: University of Nebraska Press, 2009.

Rashkin, Elissa. "La arqueología de Estridentópolis." In *Horizonte (1926–1927),* xxvii–xxxi. Facsimile ed. Mexico City: Fondo de Cultura Económica/Instituto Nacional de Bellas Artes/Gobierno del Estado de Veracruz/Universidad Veracruzana, 2011.

———. *The Stridentist Movement in Mexico: The Avant-Garde and Cultural Change in the 1920s.* Lanham MD: Lexington Books, 2009.

Richmond, Douglas W. *Venustiano Carranza's Nationalist Struggle, 1893–1920.* Lincoln: University of Nebraska Press, 1984.

Rivas Hernández, Ignacio. "La industria." In *Historia general de Baja California Sur,* volume 1, edited by Edith Gonzáles Cruz, 287–326. La Paz: Universidad Autónoma de Baja California Sur, 2002.

Rodgers, Daniel T. *Atlantic Crossings: Social Politics in a Progressive Age.* Cambridge MA: Belknap Press of Harvard University Press, 2000.

Rodríguez Benítez, Rigoberto. "La formación de ingenieros en el Colegio Rosales." In *Formación de ingenieros en el México del siglo XIX,* edited by María de la Paz Ramos Lara and Rigoberto Rodríguez Benítez, 131–72. Mexico City: Universidad Nacional Autónoma de México, 2007.

Rolland, Jorge M. "Estadio Xalapeño Heriberto Jara Corona." *Relatos e Historias en México* 86 (2015): 70–75.

———. "Estudios de Modesto C. Rolland." Unpublished document, 2014.

———. "Ingeniero Modesto C. Rolland: Pionero del uso de prefabricado de cemento." *Construcción y Tecnología en Concreto* 6, no. 11 (2015): 36–41.

———. "La monumental Plaza de Toros México: Setenta años de un portento arquitectónico." *Relatos e Historias en México* 93 (2016): 64–69.

———. *Modesto C. Rolland: Constructor de México moderno.* La Paz: Gobierno del Estado de Baja California Sur and Universidad Autónoma de Baja California Sur, 2017.

———. "Modesto C. Rolland: Vida y obra." Unpublished document, 2011.

———. "Un constructor del México moderno: Modesto C. Rolland (1881–1965)." *Relatos e Historias en México* 83 (2015): 78–84.

Rolland, Modesto C. "The Agrarian Problem in Mexico." Address delivered at the Henry George Foundation of America Congress, San Francisco, California, September 1930 (reprinted from *Land and Freedom*, November–December 1930). Available at Cooperative Individualism, accessed May 16, 2011. http://www.cooperativeindividualism.org/rolland-modesto_agrarian-problems-in-mexico.html.

———. *The Agrarian Question and Practical Means of Solving the Problem*. Veracruz: Compañía Veracruzana de Publicidad, 1914.

———. *Algunas lecciones sobre el levantamiento de polígonos por "deflexiones."* 1906. 2nd ed. Mérida, 1916.

———. *Carta a mis conciudadanos*. New York: Latin-American News Association, 1917.

———. *Cemento Armado: Elementos de cálculo*. 3rd ed. Mexico City: Talleres Gráficos "marte" Justo Sierra 67, 1948.

———. *Comunismo o liberalismo?* 2nd ed. Mexico City: publisher unknown, 1939.

———. *Distribución de las tierras: Estudio sobre Nueva Zeelandia, utilidad de la lección para México*. New York: publisher unknown, 1914.

———. *Efectiva manera de evitar la miseria pública y combatir al comunismo*. Mexico City: publisher unknown, 1952.

———. *El desastre municipal en la República Mexicana*. 2nd ed. Mexico City: I. Molina M., 1939.

———. "Free Ports and the Interoceanic Traffic." Unpublished paper presented at the Mexican Geographic and Statistical Society conference, November 5, 1940.

———. *Informe sobre el Distrito Norte de la Baja California*. Mexicali: Universidad Autónoma de Baja California, 1993.

———. "Investigation Work into the Municipal City Governments and the Rural School System, Factories and Industrial Centres in the United States." In *Carranza and Mexico*, by Carlo di Fornaro. New York: Mitchell Kennerly, 1915.

———. *Jalapa-Enríquez: Sus obras; La Universidad Veracruzana, el Estadio, la Ciudad Jardín*. Xalapa: Gobierno del Estado Libre y Soberano de Veracruz Llave, 1925.

———. "Labor Law of Yucatan." In *International Labor Forum*, edited by Modesto C. Rolland. New York: Latin-American News Association, 1916.

———. *Lecciones sobre presas: Dadas en la clase de topografía, drenaje y riegos en la Escuela Nacional de Agricultura y Veterinaria*. Mexico City: publisher unknown, 1906.

———. *Los puertos libres mexicanos y la zona libre en la frontera de la República Mexicana*. Mexico City: Empresa Editorial de Ingenería y Arquitectura, 1924.

———. "Open Letter of Mr. M. C. Rolland to Mr. W. R. Hearst." Press release, Latin-American News Association, 1916.

———. *Overland Ship Transportation across the Isthmus of Tehuantepec.* Mexico City: publisher unknown, 1951.

———. "Petroleum in Mexico." *International Socialist Review* 17, no. 3 (1917): 149–53.

———. *Problema de la Baja California.* N.p, ca. 1916.

———. *A Reconstructive Policy in Mexico.* New York: Latin-American News Association, 1917.

———. *Salina Cruz y rehabilitación del Istmo de Tehuantepec.* Mexico City: publisher unknown, 1951.

———. *Salvemos la patria: Impuesto único.* New York: Latin-American News Association, ca. 1916.

———. *Stationary Dredge.* Salina Cruz: Mexican Free Ports, 1950.

———. *Transportation of Boats across the Isthmus of Tehuantepec.* Mexico City: publisher unknown, 1946.

———. *Transporte de buques por el Istmo de Tehuantepec.* Mexico City: publisher unknown, 1946.

———. "A Trial of Socialism in Mexico: What the Mexicans Are Fighting For." *Forum* (July 1916): 79–90.

———. "Why Is a Government Needed in Mexico?" In *Mexican Problems*, by Robert Bruce Brinsmade and Modesto C. Rolland. New York: publisher unknown, 1916.

———. "Women in Mexico." In *Mexican Problems*, by Robert Bruce Brinsmade and Modesto C. Rolland. New York: publisher unknown, 1916.

Rolland, Modesto C., and Ulises Irigoyen. *Justificación del establecimiento y explotación de los puertos libres mexicanos y del tráfico interoceánico: Plan integral del Istmo de Tehuantepec.* Mexico City: Puertos Libres Mexicanos, 1941.

Román Kalisch, Manuel Arturo. "Desarrollo de tecnología constructiva en las viviendas en serie meridana del siglo XX." In *Evaluación de la vivienda construida en serie urbana arquitectónica en los desarrollos habitaciones*, edited by María Elena Torres Pérez, 69–87. Mérida: Universidad Autónoma de Yucatán, 2014.

Rosenberg, Emily S. *Spreading the American Dream: American Economic and Cultural Expansion, 1890–1945.* New York: Hill and Wang, 1982.

Ross, Stanley R. *Francisco I. Madero: Apostle of Democracy.* New York: Columbia University Press, 1955.

———. *Is the Mexican Revolution Dead?* 2nd ed. Philadelphia: Temple University Press, 1975.

Rowe, Leo S., ed. *The Purposes and Ideals of the Mexican Revolution.* Philadelphia: American Academy of Political and Social Science, 1917.

Ruíz, Ramón Eduardo. *The Great Rebellion: Mexico, 1905–1924*. New York: Norton, 1982.

——— . *Mexico: Why a Few Are Rich and the People Poor*. Berkeley: University of California Press, 2010.

Ruiz Fernández, Ana Carolina, et al. "Effects of Land Use Change and Sediment Mobilization on Coastal Contamination (Coatzacoalcos River, Mexico)." *Continental Shelf Research* 37, no. 2 (2012): 57–65.

Sáenz, Olga. *El símbolo y la acción: Vida y obra de Gerardo Murillo, Dr. Atl*. Mexico City: El Colegio Nacional, 2005.

Saldaña, Juan José. *Las revoluciones políticas y la ciencia en México*. 2 vols. Mexico City: CONACYT, 2010.

Sánchez Barría, Felipe. "El funcionamiento de los ferrocarriles en México durante los primeros años de la Revolución, 1911–1914." *Tiempo y Espacio* 21, no. 24 (2010): 73–82.

Sanderson, Steven E. *Agrarian Populism and the Mexican State: The Struggle for Land in Sonora*. Berkeley: University of California Press, 1981.

Saragoza, Alex M. *The Monterrey Elite and the Mexican State, 1880–1940*. Austin: University of Texas Press, 1990.

Schaffer, Daniel. *Garden Cities for America: The Radburn Experience*. Philadelphia: Temple University Press, 1982.

Schantz, Eric Michael. "All Night at the Owl: The Social and Political Relations of Mexicali's Red-Light District, 1913–1925." Special issue, "Border Cities and Culture," *Journal of the Southwest* 43, no. 4 (2001): 549–602.

Scheips, Paul J. "Gabriel Lafond and Ambrose W. Thompson: Neglected Isthmian Promoters." *Hispanic American Historical Review* 36, no. 2 (1956): 211–28.

Schivelbusch, Wolfgang. *The Railway Journey: The Industrialization of Time and Space in the 19th Century*. Berkeley: University of California Press, 1986.

Schrecker, John E. *The Chinese Revolution in Historical Perspective*. Westport CT: Praeger, 2004.

Schuler, Friedrich. *Mexico between Hitler and Roosevelt: Mexican Foreign Relations in the Age of Lázaro Cárdenas, 1934–1940*. Albuquerque: University of New Mexico Press, 1999.

——— . *Secret Wars and Secret Policies in the Americas, 1842–1929*. Albuquerque: University of New Mexico Press, 2010.

Scott, James C. *The Art of Not Being Governed: An Anarchist History of Upland Southeast Asia*. New Haven: Yale University Press, 2009.

——— . *Seeing Like a State: How Certain Schemes to Improve the Human Condition Have Failed*. New Haven: Yale University Press, 1997.

——— . "State Simplifications: Nature, Space, and People." *Nomos* 38 (1996): 42–85.

Secrest, Meryle. *Frank Lloyd Wright: A Biography*. Chicago: University of Chicago, 1998.

Secretaria de Comunicaciones y Obras Públicas. *Ferrocarril del Sureste*. Mexico City: Talleres Gráficos de la Nación, 1950.

Shadle, Stanley F. *Andrés Molina Enríquez: Mexican Land Reformer of the Revolutionary Era*. Tucson: University of Arizona Press, 1994.

Slaton, Amy. *Reinforced Concrete and the Modernization of American Building, 1900–1930*. Baltimore: Johns Hopkins University Press, 2001.

Smith, Benjamin T. "Building a State on the Cheap: Taxation, Social Movements, and Politics." In *Dictablanda: Politics, Work, and Culture in Mexico, 1938–1968*, edited by Paul Gillingham and Benjamin T. Smith, 255–76. Durham: Duke University Press, 2014.

Smith, Michael M. "Andrés G. Garcia: Venustiano Carranza's Eyes, Ears, and Voice on the Border." *Mexican Studies/Estudios Mexicanos* 23, no. 2 (2007): 355–86.

———. "Carrancista Propaganda and the Print Media in the United States: An Overview of Institutions." *The Americas* 52, no. 2 (1995): 155–74.

———. "The Mexican Secret Service in the United States, 1910–1920." *The Americas* 59, no. 1 (2002): 65–85.

Smith, Peter. "The Making of the Mexican Constitution." In *The History of Parliamentary Behavior*, edited by William O. Aydelotte, 186–224. Princeton: Princeton University Press, 1977.

———. "The Mexican Revolution and the Transformation of Political Elites." *Boletín de Estudios Latinoamericanos y del Caribe* 25 (December 1978): 3–20.

Smith, Stephanie. "Removing the Yoke of Tradition: Yucatán's Revolutionary Women, Revolutionary Reforms." In *Peripheral Visions: Politics, Society, and the Challenges of Modernity in Yucatan*, edited by Edward D. Terry, Ben W. Fallaw, Gilbert M. Joseph, and Edward H. Mosely, 79–100. Tuscaloosa: University of Alabama Press, 2010.

———. "Salvador Alvarado of Yucatán: Revolutionary Reforms, Revolutionary Women." In *State Governors in the Mexican Revolution: Portraits in Conflict, Courage, and Corruption*, edited by Jürgen Buchenau and William H. Beezley, 43–58. Lanham MD: Rowman and Littlefield, 2009.

Spencer, Daniela, and Richard Stoller. "Radical Mexico: Limits to the Impact of Soviet Communism." Special issue, "Reassessing the History of Latin American Communism," *Latin American Perspectives* 35, no. 2 (2008): 57–70.

Stabile, Donald R. "Valben and the Political Economy of the Engineer: The Radical Thinker and Engineering Leaders Came to Technocratic Ideas at the Same Time." *American Journal of Economics and Sociology* 45, no. 1 (1986): 41–52.

Stein, Harry H. "Lincoln Steffens and the Mexican Revolution." *American Journal of Economics and Sociology* 34, no. 2 (1975): 197–212.

Steinbeck, John. *The Log from the Sea of Cortez*. 1941. New York: Penguin Classics, 1995.

——— . *The Pearl*. 1947. New York: Penguin Books, 1992.

Steinbeck, John, and Rosa Harvan Kline. *The Forgotten Village: Life in a Mexican Village*. 1941. New York: Penguin Books, 2009.

Stout, Joseph A., Jr. *Border Conflict: Villistas, Carrancistas and the Punitive Expedition, 1915–1920*. Fort Worth: Texas Christian University Press, 1999.

Sutcliffe, Anthony, ed. *Metropolis, 1890–1940*. Chicago: University of Chicago Press, 1984.

——— , ed. *The Rise of Modern Urban Planning, 1800–1914*. New York: Spon Press, 1998.

Tafunell, Xavier. "On the Origins of ISI: The Latin American Cement Industry, 1900–1930. *Journal of Latin American Studies* 39, no. 2 (2007): 299–328.

Tamayo, Jorge L. *Breve reseña sobre la Escuela Nacional de Ingeniería*. Mexico City: Imprenta La Esfera, 1958.

Tannenbaum, Frank. *The Mexican Agrarian Revolution*. New York: Macmillan, 1929.

——— . *Peace by Revolution: An Interpretation of Mexico*. New York: Columbia University Press, 1933.

Taracena, Alfonso. *Venustiano Carranza*. Mexico City: Editorial Jus, 1963.

Tenorio-Trillo, Mauricio. "The Cosmopolitan Mexican Summer, 1920–1949." *Latin American Research Review* 32, no. 3 (1997): 224–42.

——— . *I Speak of the City: Mexico City at the Turn of the Twentieth Century*. Chicago: University of Chicago Press, 2013.

——— . "1910 Mexico City: Space and Nation in the City of the Centenario." *Journal of Latin American Studies* 28, no. 1 (1996): 75–104.

——— . "Stereophonic Scientific Modernisms: Social Science between Mexico and the United States, 1880s–1930s." Special issue, "A Nation and Beyond: Transnational Perspectives on United States History," *Journal of American History* 86, no. 3 (1999): 1156–87.

Thoman, Richard S. *Free Ports and Foreign Trade Zones*. Cambridge MD: Cornell Maritime Press, 1956.

Tibol, Raquel. *Frida Kahlo: An Open Life*. Translated by Elinor Randall. Albuquerque: University of New Mexico Press, 1993.

Tinajero, Araceli, and J. Brian Freeman, eds. *Technology and Culture in Twentieth-Century Mexico*. Tuscaloosa: University of Alabama Press, 2013.

Tobin, Eugene M. "The Insurgent as Ideologue: George L. Record and the Single Tax in Mexico; A Research Note." *American Journal of Economics and Sociology* 35, no. 3 (1976): 325–31.

Toth, Charles. "Bulwark for Freedom: Samuel Gompers' Pan American Federation of Labor." Monograph supplement, *Revista/Review Interamericana* 9, no. 3 (1979): 457–91.

Trumpbour, Robert C. *The New Cathedrals: Politics and Media in the History of Stadium Construction*. Syracuse NY: Syracuse University Press, 2007.

Tyrrell, Ian. *True Gardens of the Gods: Californian-Australian Environmental Reform, 1860–1930*. Berkeley: University of California Press, 1999.

Ulloa, Berta. *Historia de la Revolución Mexicana, 1914–1917*. Mexico City: El Colegio de México, 1979.

U.S. Congress. Senate Committee on Foreign Relations. *Investigation of Mexican Affairs*. 2 vols. Washington DC: Government Printing Office, 1920.

U.S. State Department. *Papers Relating to the Foreign Relations of the United States with the Annual Message of the President Transmitted to Congress December 3, 1912*. Washington DC: U.S. Government Printing Office, 1912.

Vanderwood, Paul. *The Power of God against the Guns of Government: Religious Upheaval in Mexico at the Turn of the Nineteenth Century*. Stanford: Stanford University Press, 1998.

———. *Satan's Playground: Mobsters and Movie Stars at America's Greatest Gaming Resort*. Durham: Duke University Press, 2010.

Vaughn, Mary Kay. *State, Education, and Social Class in Mexico, 1880–1928*. DeKalb: Northern Illinois University Press, 1982.

Vázquez Schiaffino, José. "Memoria relativa al viaje efectuado a los Estados Unidos de América, por una parte del personal de la Comisión Técnica del Petróleo por J. Vázquez Schiaffino, Ing. Civil." *Boletín de Petróleo* 2, no. 6 (1916): 506–34.

Velázquez Estrada, Rosalía, ed. *Salvador Alvarado*. Mexico City: Instituto Nacional de Estudios Históricos de la Revolución Mexicana, 1985.

Wakild, Emily. "Naturalizing Modernity: Urban Parks, Public Gardens and Drainage Projects in Porfirian Mexico City." *Mexican Studies/Estudios Mexicanos* 23, no. 1 (2007): 101–23.

Warfield, Adrienne Akins. "Steinbeck and the Tragedy of Progress." In *A Political Companion to John Steinbeck*, edited by Cyrus Ernesto Zirakzadeh and Simon Stow, 98–117. Lexington: University Press of Kentucky, 2013.

Wasserman, Mark. *Pesos and Politics: Business, Elites, Foreigners, and Government in Mexico, 1854–1940*. Stanford: Stanford University Press, 2015.

Weis, Robert. "The Revolution on Trial: Assassination, Christianity, and the Rule of Law in 1920s Mexico." *Hispanic American Historical Review* 96, no. 2 (2016): 320–53.

Weitz, Eric D. *Weimar Germany: Promise and Tragedy*. Princeton: Princeton University Press, 2007.

Winberry, John J. "The Mexican Landbridge Project: The Isthmus of Tehuantepec and Inter-Oceanic Travel." *Yearbook: Conference of Latin Americanist Geographers* 13 (1987): 12–18.

Wolfe, Mikael. "Bringing the Revolution to the Dam Site: How Technology, Labor, and Nature Converged in the Microcosm of a Northern Mexican Company Town, 1936–1946." *Journal of the Southwest* 53, no. 1 (2011): 1–31.

———. *Watering the Revolution: An Environmental and Technological History of Agrarian Reform in Mexico*. Durham: Duke University Press, 2017.

Wood, Andrew Grant. "Adalberto Tejada of Veracruz: Radicalism and Reaction." In *State Governors in the Mexican Revolution, 1910–1952: Portraits in Conflict, Courage, and Corruption*, edited by Jürgen Buchenau and William H. Beezley, 77–94. Lanham MD: Rowman and Littlefield, 2009.

Womack, John, Jr. *Zapata and the Mexican Revolution*. New York: Vintage, 1970.

Wukovitz, John F., ed. *The 1910s*. San Diego: Greenhaven Press, 2000.

Yankelevich, Pablo. "En la retaguardia de la Revolución Mexicana: Propaganda y propagandistas mexicanos en América Latina, 1914–1920." *Mexican Studies/Estudios Mexicanos* 15, no. 1 (1999): 35–71.

Yardley, Edmund, ed. *Addresses at the Funeral of Henry George*. Chicago: Public Publishing Company, 1905.

Zapata Vela, Carlos. *Conversaciones con Heriberto Jara*. Mexico City: Costa-Amic Editores, 1992.

Index

ABC Commission, 47
Abrahan, José, 223
Addams, Jane, 87
Aero-Motor México, 164–65, 191
Agrarian Decree (of January 6, 1915), 51, 57–58, 70, 145
Agrupación Democrática Pacificadora Nacional. *See* National Democratic Peacemaker Group
Agua Prieta Revolt, 40, 116–19, 126, 137
Aguilar, Francisco J., 188
Alemán Valdés, Miguel, 198, 210
Alvarado, Salvador: and the Agua Prieta Revolt, 118–19; and the De la Huerta Revolt, 137; and *El Heraldo de México*, 111; as governor of Yucatán, xviii, 58, 60, 65–77, 80–81, 84, 98–100, 104–5, 109–10, 124, 142–43; and the military invasion of Yucatán, 65, 142; and national civil service law proposals, 129; as potential presidential candidate, 100, 111–12, 117–18; and the Revolutionary Confederation, 56, 65; Rolland's promotion of, 67, 87–88, 99–100, 105, 111, 117; as secretary of finance, 119, 123–26; and Tehuantepec, 109, 119, 123–26, 241
American Academy of Political and Social Science, 91–93
American Federation of Labor, 71
American Union against Militarism, 82–83, 87–90, 101
Angulo, Melquiades, 188–89
Anti-Imperialist League, 89
Anti-Reelectionists, 19–20, 22–26
Antonio Alzante Scientific Society, 8
Anza, Antonio M., 15, 233

apoliticism, xviii–xix, 11, 17, 84, 122, 175–76, 189
Arango, Doroteo. *See* Villa, Pancho
Austin, Mary Hunter, 101
Ávila Camacho, Manuel, 189–90, 196–210, 215–16, 219
Azcárraga, Emilio, 134
Azcárraga, Luis, 134
Azcárraga, Raúl, 134

Baja California (Northern District of Baja California and Northern Territory of Baja California): Constitutionalist commissions in, 105, 113–16; and Esteban Cantú, 79–80, 113–16, 111–12; and the Foreign Club, 163; and free zones, 191–92; Modesto Rolland's idea for a canal across, 77–79; and Ricardo Flores Magón, 33
Baja California peninsula: and Adolfo López Mateos's presidential campaign, 208; as compared to the Yucatán peninsula, 66; and Francisco Madero, 35; as an influence on Modesto Rolland's ideas, 26, 28, 62, 77–81, 104, 108–9, 111–12, 165, 183, 190–93; and John Steinbeck, xiii–xv, 247nn1–2; and Lázaro Cárdenas, 190–92; as a military zone, 205; and the National Agrarian Commission, 126; organizations, 22–23, 27–28, 35, 105, 191; political organization of, 5–6, 162
Baja California Sur (Southern District of Baja California and Southern Territory of California): and free zones, 191–92; and Modesto Rolland's childhood, xiii, 1–6, 245; Modesto Rolland's legacy in, 239; Modesto Rolland's nomination as congressional representative of, 104

Balcárcel, Blas, 16
Banco de México, 159, 175
Bandala Fernández, Eliseo, 232
Barragán, Juan, 186
Baumeister, Reinhard, 10
Beauregard, Paul, 10
Bellamy, Edward, 8, 146
Beteta, Ramón, 223
Birkmire, William H., 10
Bonillas, Ignacio, 92–94, 116–18
Boone, William K., 150, 238
Boyd, Charles T., 85–86
Brown, Edward N., 19, 30–31, 51
Bryan, William Jennings, 89
Bucareli Accords, 139
Buen Abad, Manuel, 204
Bulnes, Francisco, 233, 241
Burns, Daniel E., 97

Cabrera, Federico, 16
Cabrera, Luis: and agrarian reform, 51, 57–58, 77; as Anti-Reelectionist, 19; as Carranza adviser and official, 43, 51, 57–58, 92–93, 109, 124; and disillusionment with the Mexican Revolution, 221; and the Inter-American Peace Committee, 92–93; and Tehuantepec, 109, 124, 241
Cabrera, Mariano, 124, 128
California Oil and Asphalt Company, 64
Calles, Plutarco Elías: and the Agua Prieta Revolt, 118; as Constitutionalist commander in Agua Prieta, 84; and the De la Huerta Rebellion, 136–37; and Heriberto Jara, 142, 151–53, 155; and the inauguration of the Xalapa stadium, 152–53; and Lázaro Cárdenas, 178, 180, 182; and the Maximato, 173, 178, 241; and Modesto Rolland, xviii, 142–43, 158, 165, 174–75, 182; presidency of, 142–43, 151, 158–61, 174–75, 241; presidential cabinet of, 141; presidential inauguration of, 150
Cantú, Esteban, 79–80, 111–16
Cárdenas, Lázaro, xviii, 175–93, 196–97, 202, 204, 232, 242
Cardineault, Edmundo, 15–16, 128
Carranza, Venustiano: and agrarian reform, xviii, 51, 56–58, 70, 76, 110, 131, 145; and the Agua Prieta Revolt, 40, 116–19; and the Constitution of 1917, 103–4; death of, 116, 119, 123; and education, 42–43; and foreign policy with Latin America, 63; and foreign policy with the United States, xxi–xxii, 40, 42–43, 47–50, 58, 63–65, 67, 83–87, 89–94, 99–100, 104; and Francisco Pendás's conflict with Modesto Rolland, 98–100; as governor of Coahuila, 41; as head of Constitutionalists, xvi, xviii, xxii, 40–42, 46–50; Modesto Rolland's criticism of, 56, 58, 106, 138; and nationalism, 42, 45, 83; as opponent of Ricardo Flores Magón, 97; and the petroleum commission, 58, 63–65; and progressive reform, 45, 74; split with Pancho Villa, 47, 49–50, 55, 80, 84; and taxes and fees, 68; and technocrats, 42–43, 50, 58–59, 80, 89, 92, 109, 117, 131; and territorial control, 46, 68, 79–80, 104, 111, 114; and the Veracruz conference of 1914–15, 55–58, 145; and Yucatán, 65, 70, 72, 75, 105, 109–10
Carrizal: battle or incident in, 85–86, 88–89, 91, 237
Carvajal, Francisco, 48
Caso, Alfonso, 128
Castro, Fidel, 230
Castro, Luis "El Soldado," 207
Castro Morales, Carlos, 105, 110
Catholic Church: landholdings of, 52, 108, 218–19; Rolland's relationship with, 81; state conflicts with, 27, 75; and women, 75
cement, 15–17, 159
Central Mexican Radio League (LCMR), 133, 138
Central Radio Club, 133
Centro de Ingenieros. See Engineers' Center
Chávez, Ezequiel A., 129
Chernyshevsky, Nikolai, 156
Científicos, xx, 4, 23, 46, 48
City of Sports, 198–99, 205–8, 215, 229, 231, 239
Clientelism, xviii, 158, 170–72
Club de Estudios Económicos-Sociales. *See* Social-Economic Studies Club
Cockcroft, James. D., 171
Colegio Militar, 11, 36, 38, 207

Comisión del Suchiate. *See* Suchiate Commission
communism, 169–71, 174–79, 181, 184, 217, 238
Compañia Mexicana de Petróleos, 185
Compañia Petrolera Veracruzana, 185–87, 197
concrete, xvi, 13, 15–17, 22, 26, 110, 115, 141–42, 149–51, 153, 155, 212, 214, 231, 236–37
Confederación Regional Obrera Mexicana. *See* Regional Confederation of Mexican Workers (CROM)
Congreso Nacional Agronómico. *See* National Agronomic Congress
Congreso Nacional de Exportadores. *See* National Congress of Exporters
Cortés, Hernán, 123
Cosío Villegas, Daniel, 168, 221
Creelman, James, 17
Cristero Revolt, 155, 160, 161
Cuba, 61
CYB, radio station, 134
CYL, radio station, 134

Daniels, Josephus, 187
Decena Trágica. *See* Ten Tragic Days
de Fornaro, Carlo. *See* di Fornaro, Carlo
de Fuentes, Fernando, 113
de Garay, Francisco, 16
de la Barra y Quijano, Francisco León, 27–28
De la Garza Meléndez, Virginia. *See* Garza de Rolland, Virginia
de la Garza Velazco, Miguel, 13–14
de la Huerta, Adolfo: and the Agua Prieta Revolt, 118–19; and the De la Huerta Rebellion, 136–39, 142, 145; as minister of finance, 168; as provisional president, 116, 119, 121–24, 126–27, 159; and the Veracruz conference (1914–15), 56
De la Huerta Rebellion, 136–39, 142, 145
de la Luz Blanco, José, 25
Del Castillo, Luz Elvira, 12
de León Toral, José, 160
de Lesseps, Ferdinand, 209
de Quevedo, Miguel Ángel, 16, 192, 233
Desagüe, 8, 16

Development Company of the Southeast, 72
Dewey, John, 43, 206, 220
Díaz, Félix, 34–37, 104
Díaz, Porfirio: advisors to, 4–5, 15, 17–18, 23, 46, 169, 233; agrarian policies under, 51–52; and Anti-Reelectionists, 19–20, 22–24; and Baja California, 4–6; as civilizer, 101–2; as dictator, xvi, 17, 19–20, 97; and education, 4–5, 9, 12; federal budget under, 167; and foreign capitalists, 1–2, 4, 18, 195; and inequality, 20, 55, 230; and Mexico City, 7, 16; and modernization, xvi, 4, 20; overthrow and exile of, 22, 25–26, 105, 166; proclaimed connections to Conventionalists, 50; and public works, 4, 9, 16, 21, 123–24; and railroads, 18; rebellion against, 22–26; and re-election, 19, 21–22, 24; and repression, 17; and technocrats, 4, 15, 17, 210, 223, 233; and Tehuantepec 123–24, 210, 240; U.S. sympathizers of, 32, 96–97
Díaz Covarrubias, Francisco, 10
di Fornaro, Carlo, 37–38, 48–51, 61, 67, 88, 99
divorce, 14, 16, 74, 92
Dolores water tanks, 16
Domínguez, Norberto, 129
Dr. Atl, 56, 58, 88–90

Eads, James, 209–10
Eastman, Crystal, 87–88
education: and Alberto Pani, 94; and Baja California, 5, 22, 27–28, 104; and cities, 128; and the Constitutionalists, 94, 102; and Eliseo Bandala Fernández, 232; and engineering, 8–12, 15, 234; and Esteban Cantú, 114–16; in Europe, 45; during the Francisco León de la Barra administration, 22, 27–28; of indigenous peoples in Mexico, 54, 67, 169; and infrastructure, 180; and John Dewey, 43, 206; and John Steinbeck, xiv, 247n2; and José Vasconcelos, 149–50; during the Lázaro Cárdenas, 202; during the Manuel Ávila Camacho administration, 206; and Modesto Rolland, xv–xvi, 2, 5–12, 27, 41–43, 45, 54, 66–67, 73, 77, 80, 104, 108,

education (*continued*)
133–34, 154–56, 218, 237, 243, 245; in New Zealand, 45; and piano, 162–63; and Plutarco Elías Calles, 152; during the Porfirian era, 4–13, 22, 67, 234; and radio, 122, 133; in the United States, 43, 45, 91–92; U.S.-Mexican exchanges in, 91–93, 245; and Virginia de la Garza Meléndez, 12–13; and Venustiano Carranza, 42–43; and women, 12–13, 73–74; in the Yucatán, 54, 66, 73–74, 76, 104

Einstein, Albert, 220

ejidos, 57–58, 70, 127, 129, 131, 137, 166–69, 174, 178, 180, 219, 230

El Boleo, 5

El Buen Tono, 7, 134

El Gráfico, 88, 99–100, 106

El Heraldo de México, 111–13, 117–18, 120, 122–23

Ellis, William H., 124

engineers: and agrarian reform policies, 52, 66; as Anti-Reelectionists, 19–20; as apolitical, xviii–xix, 17, 122; British, 16, 123–24; and built environments, xix, 155–57; and city planning, 15, 112, 130; and civil service legislation, 129; and class conciliation, 180; and the Constitution of 1917, 103–4; and desire to control, 11, 15, 226–27; as diplomats, 43, 92, 101; disillusionment of, 221; Dutch, 214; and education, 7–12, 15, 235, 234; and the free ports, 202, 204, 220, 223, 225, 227; French, 209; and globalization, 234; limited number of, 105, 121, 204; and Mexicanization, 18, 22, 26–29; and military instruction, 36; in Modesto Rolland's Garden City in Xalapa, 147; Modesto Rolland's generation of, 11, 16–18, 26, 37, 43, 112, 127–28, 130, 221, 234; and municipal governance, 129–30; and the National Agrarian Commission, 126–27; and nation-state consolidation, 6, 26–27, 202; and politics, xix, 101, 105–6, 112–13, 122, 126, 232; as polymaths, 101, 235; and postal service, 18; and the Pro-California Commission, 192; and the Pro-Tree Committee, 192; and radio, 133; and railroads, 18, 42, 182, 204, 209–10, 213; as revolutionaries, 24–25, 37; and roads, 130; and sanitation, 11, 16, 94, 112; and the Social-Economic Studies Club, 127–28; and social unification, 180; and stadiums, xix, 149, 151, 155; and the stationary dredge, 214–15, 225; and the Suchiate Commission, 202–4; as systems and network builders, xvii, 15, 42; and technocracy, xv, xvii, xix, xxii, 52, 105–6, 180, 221, 233–34, 242; and Tehuantepec, 123–24, 204, 209–10, 241; and top-down planning, 35, 226–27, 242; and transisthmian projects, 209; and turnover in government positions, 143; and unionization, 130; U.S., 209, 213–14, 221; and water projects, 16; and wealth, 112;

Engineers' Center, 112–13, 127, 130, 133, 138

Engineers' Club, 18, 27

environmental problems, 166–67, 183, 226–27, 242–45

environmental protections, 183, 192, 243

Erie Canal, 16

Escobar, José Gonzalo, 161, 227

Escobarista Rebellion, 227

Estadio Azul. *See* Olympic Stadium

Estadio Heriberto Jara Corona. *See* Xalapa Stadium

Estadio Nacional. *See* National Stadium

Estadio Olímpico. *See* Olympic Stadium

Estadio Xalapeño. *See* Xalapa Stadium

Estrada, Enrique, 131

Estrada Reynoso, Roque, 128

Estridentista. *See* Stridentist

Estridentópolis, 141

Executive Committee of Ejido Administration, 131

Félix, Carlos A., 134

Ferrocarril del Sureste. *See* Southeastern Railroad

Ferrocarriles Nacionales de México. *See* National Railways

fertilizers, 78, 164, 190, 201, 218–19

First National Congress on Roads, 130

Fisher, Irving, 90

fixed dredge. *See* stationary dredge

Flores, Trinidad, 118

Flores Magón, Ricardo, 33, 97

Foreign Club, 163, 231

Fourier, Charles, 8

France, 82–83
Franco, Francisco, 179
free ports: during the Adolfo de la Huerta administration 119, 123, 125–26, 134; during the Adolfo Ruiz Cortines administration, 222–27; Álvaro Obregón's closure of, 139, 142, 158, 174, 186, 241; debates about, 120, 125–26, 174, 204, 224–25, 241; definition of, 124–25; establishment of, 119, 123–24, 126, 135, 138–39; failure of, 217, 220–21, 233–34, 239–41; history of 124; during the Lázaro Cárdenas, 188–91, 193, 204; during the Manuel Ávila Camacho administration, 198–200, 202, 206, 210; during the Miguel Alemán Valdes administration, 198, 212; Modesto Rolland's management of, 119, 121, 123, 129, 132, 134–35, 139, 180, 188–89, 197, 199–200, 202, 204–5, 210–13, 215, 222, 227, 231, 239; in New Orleans, 213; policies, 125–26, 225–26; promotion of, 162, 188, 194, 213; ships for the, 135; in Staten Island, 213, 224; success of, 242–43
Free Ports Commission, 121, 123, 174, 190, 200, 213

Gamio, Manuel, 169
garden city, 145–48, 155–57, 206, 267n25
Garner, Paul, 240
Garza, Lorenzo, 207–8
Garza Aldape, Manuel, 39, 41
Garza de Rolland, Virginia, 12–15, 22, 39, 74, 98, 136–37, 163
Gauss, Susan, 168
Gayol y Soto, Roberto, 16
Geological Institute, 64
George, Henry, 54, 61, 67–68, 81, 122, 148, 156, 158, 169; policies influenced by, 61, 69, 72, 99, 108, 116, 238; publications influenced by, 128, 146, 153, 238
globalization, xv, 243
Glorieta de Colón, 16
Gómez, Arnulfo, 155
Gómez, Efraín R., 37
Gómez, Félix T., 86
Gómez Morín, Manuel, 127–28, 159, 175, 221
Gompers, Samuel, 71, 111
González, Manuel, 1–2, 4, 210

González, Pablo, 46, 49, 117
Grand Radio Fair, 134, 140–41
Gray, George, 92
Great Depression, 170, 179–80
Gruson, Sydney, 222
Gurza, Jaime, 31
Gutiérrez Ortiz, Eulalio, 49
Gúzman, Martín Luis, 137

Hall, Peter, 147
Hammeken, Carlos M., 124
Hay, Eduardo, 128
Hearst, William Randolph, 86–87, 89–90, 92–94
Henry George Foundation, 166
Hernédez Palacios, Aureliano, 141
Hidalgo: state of, 15–38
Hidalgo, Miguel, 21
Hitler, Adolf, 177–78, 184
Holmes, John Haynes, 96
Hotel Chula Vista, 164, 231
Howard, Ebenezer, 146–48, 156
Howe, Jerome W., 85
Huerta, Victoriano, 34, 37–41, 46–48, 50, 52

important substitution industrialization (ISI), 198
indigenismo, 67
inequality, 20, 27, 51, 54, 171, 208, 219, 230, 244
Infante, Pedro, 232
Inter-American Peace Committee, 92–93
Irigoyen, Ulises, 192, 194
irrigation, xix, 11, 27, 52, 92, 108, 111, 162, 165, 191, 203, 218–19, 252n16
Iturbide, Ramón F., 128
Iturriaga, José, 221

Jalapa Railroad and Power Company, 150
Japan: diplomats from 152; immigrants from, 79; naval commander of, 64; oil interests of, 178, 184–88, 194, 196, 199; and the rayon trade, 196–97; rightest government of, 179; U.S. war with, 200, 205
Jara Corona, Heriberto: in the Cárdenas administration, 182; descendants of 238; as governor of Veracruz, xviii, 141–46, 150–55, 157–58, 181; as head of the Ministry of Marina, 204; as military commander, 65, 112, 143

Johnson, Jack, 62
Joint Peace Commission. *See* Mexican-American Peace Commission
Jordan, David Starr, 89–91, 96
Joseph, Gilbert, 76
Josephson, Paul R., xviii
Juárez, Benito, 9

Kahlo, Frida, 177–78, 184
Kansas City Railroad, 190
Kellogg, Paul, 90
Kloppenberg, James T., xxi
Klor, Charles, 224
Knight, Alan, 24
Knox, Philander C., 32
Kropotkin, Peter, 10

La Casa de Radio, 134
Lane, Franklin K., 92
Las Minas, 201
Latin-American News Association, 67, 88, 92, 95, 98–99, 101, 104, 106
Law of Unused Lands (1920), 127, 129
Lazo Barreiro, Carlos, 227–28;
Lea, Tom, 89
Le Corbusier, 156
Lenin, Vladimir I., 181
Ley de Tierras Ociosas. *See* Law of Unused Lands (1920)
Liberal Constitutional Party. *See* Partido Liberal Constitucional
liberalism, xix–xxi, 20
Liga Central Mexicana de Radio. *See* Central Mexican Radio League (LCMR)
Limantour, José Y., 18, 23, 233
Lineas Férreas de México. *See* Railway Lines of Mexico
List Arzubide, Germán, 153–54
Lombardo Toledano, Vicente, 128
López, Miguel, 113
López de Lara, César, 76
López Mateos, Adolfo, 229
Lord Cowdray. *See* Pearson, Weetman

Madero, Ernesto, 31
Madero, Francisco I., xvi, 19, 22–40, 48, 50, 52
Madero, Gustavo, 30
Magaña, Mardonio, 232
Maples Arce, Manuel, 140–41, 152–53, 155

Marroquín y Rivera, Manuel, 9, 15–16, 21, 233
Martí, José, 108
Martínez, Luis María, 207
Marx, Karl, 10, 122, 170, 177, 184
Massachusetts Institute of Technology (MIT), 116
Mayoral Heredia, Manuel, 221
Maytorena, José María, 41
Mejía Altamirano, María Jesús, 2, 5–6
Meléndez Roicio, Virgina, 13–14
Méndez, Mario, 117–18
Menocal, Mario García, 62
Merriman, Mansfield, 9
Mexican-American Peace Commission, 82, 88–93
Mexican Bureau of Investigation, 48
Mexicanization, 18, 22, 26–33
Mexican Letter, 48
Mexican Shipping of the Gulf, 135
Mexican Shipping of the Pacific, 135
Meza, Carlos, 27
middle class, xxii, 22, 24–45, 102, 210
Milholland, John E., 97
Military College. *See* Colegio Militar
Millán y Alva, Rafael N., 113
Minería de Rosario, 164
Ministry of National Economy, xviii
Minturn, Gertrude, 90
Miyamoto, Yusuke, 64
Modernism, xix
modernization, xvi, 53, 65, 222, 238, 243
Molina Enríquez, Andrés, 51–52, 57, 77, 127, 131
Mondragón, Manuel, 36
Morelos, state of, 55
Morgan, Lewis H., 168–69
Morgan, William S., 146
Múgica, Francisco J., 112, 177, 180–83, 185, 187–90
municipalities: and agrarian reform, 127; and autonomy, xx, 26–27, 78, 92, 96, 103, 108, 127, 129, 170, 193; and bureaucratic expansion, 167, 202–3; and dysfunction, 129, 218, 225; facilities for, 202; and free ports, 225; governance of, 26, 42, 44, 111, 122, 128–30, 138, 146, 167, 170, 218, 234, 237; legislation on,

122, 127; Modesto Rolland's studies and writings about, xvi, 41–42, 44–45, 80, 108, 111, 122, 128–29; and public health, 44; reform of, 26–28, 92, 128, 202–3, 218, 238; and taxes, 68, 225
Murillo, Geraldo. *See* Dr. Atl
Mussolini, Benito, 177, 184

National Action Party. *See* Partido Acción Nacional (PAN)
National Agrarian Commission, 57, 121, 123, 126–27, 129–32, 139, 145, 166
National Agrarianist Party, 131
National Agriculture and Veterinary School, 11
National Agronomic Congress, 130
National Association for the Advancement of Colored People (NAACP), 90
National Congress of Exporters, 188
National Convention of Engineers, 130
National Democratic Peacemaker Group, 35–36
National Medical Institute, 8
National Radio Club, 133
National Railroad Company, 18–19
National Railways, 18, 27–33, 50, 188
National Revolutionary Party. *See* Partido Nacional Revolucionario (PNR)
National School of Engineering, 7–11, 15, 112, 141, 231
National School of Medicine, 7
National Stadium, 149–50
Naviera Mexicana del Golfo. *See* Mexican Shipping of the Gulf
Naviera Mexicana del Pacifico. *See* Mexican Shipping of the Pacific
Nefatlí Amador, Juan, 43
New York Board of Trade, 134
New Zealand, 52–54, 67
Niblo, Stephen, 184
Noguera, José Miguel, 232
Normal School for Teachers, 12
North American Free Trade Agreements (NAFTA), 241
Novo, Salvador, 177

Obregón, Álvaro: and agrarian reform, 129–32, 137, 166; and the Agua Prieta Revolt, 40, 116–19; assassination of, 157, 160–61; and the De la Huerta Rebellion, 137–38, 142; and the free ports, 134–35, 139, 142, 174, 186; as general, 34, 40, 46, 49, 84, 117; and José Vasconcelos, 149; as militia leader in Sonora, 34; and the 1921 centennial commemoration, 130; populist politics of, 158; and radio, 133–34, 137–38; re-election of, 155, 160; relationship with Salvador Alvarado, 123; and the Revolutionary Confederation, 56
O'Gorman, Juan, 177–78, 183–84, 228
Olympic Stadium, 206–7
Organization of the Federal District and Territories Law (1917), 113
Orozco, Pascual, 25, 34
Overstreet, Harry, 90

Pailles, Antonio, 227
Palavicini, Félix, 19, 83
Panama Canal, 124–25, 135, 139, 194, 200, 208–9
Pani, Alberto J., 16, 37, 56, 83, 89, 92–94, 159, 168, 175
Pani, Arturo, 16
Pani, Julio, 16.
Parkinson, Donald, 156
Parkinson, John, 156
Partido Acción Nacional (PAN), 127–28, 175
Partido de la Revolución Mexicana (PRM), 178–79, 182–83, 188
Partido Liberal Constitucional (PLC), 131–32, 137
Partido Nacional Agrarista (PNA), 131
Partido Nacional Revolucionario (PNR), 161, 165, 175, 178
Partido Revolucionario Institucional (PRI), 175, 210, 217, 222, 228–30, 233
Paz Salinas, María Emilia, 186
Peabody, George Foster, 97
Peace Committee. *See* Mexican-American Peace Committee
Pearson, Weetman, 16, 123–24, 240
Pearson and Son, 16, 124, 135
PEMEX, 73
Pendás, Francisco "Frank," 88, 98–100, 106
Pérez Castro, Lorenzo, 124
Pershing, John, 85–87, 92
personalist politics, 27, 159, 186
Petróleos Mexicanos. *See* PEMEX

Petroleum Technical Commission, 58–59, 61–65, 237
Petroleum Workers of the Mexican Republic, 223
Pillet, Julet, 10
Pinchot, Amos, 87, 90
Pino Suárez, José María, 37
Plaza de Toros, 206–8, 231, 235, 239
Pollock, Horatio, 146
Portes Gil, Emilio, 161, 171
Positivism, xix, 46
Primer Congreso Nacional de Caminos. *See* First National Congress on Roads
Pro-California Commission, 192
Procuna, Luis, 207
Professional School of Engineers, 9
Progreso, 73
Progress and Poverty, 72
Progressives: and the Constitutionalists, xxii, 40–44, 71–72, 83–84, 88, 92–93, 101–2, 237–38; European, 44–45, 82–83, 147; influence on municipal reform, xx, 28, 44, 82–83, 129, 144; influence on political economic theory, 10; and labor reforms, 71–72, 104; taxes, 168, 220; and technocracy, 233; U.S., xxi–xxii, 40–41, 43–44, 53, 59, 67–68, 71–72, 81, 82–84, 87–88, 92, 94–95, 97, 101–2, 146, 156, 237; and World War I, 83;
progressivism: and apoliticism, xviii; and Australia, xvii, xxi, 45, 146; and China, 237; and Europe, xvii, xxi, 44–45; global, xvi, xviii, 44–45, 237; historiography of, xxi; and Mexican intellectuals, xvi, xxi, 71–72, 76, 93, 117, 122, 127–28, 156, 159, 237–39; and municipal reforms, xx; and New Zealand, xvii, xxi, 45, 52–53, 67, 146, 237; and North America, xvii, xxi, 129, 237; and socialism, 108; and South America, xvii, 23; variation within, xvii, 82–83, 129, 146–47
Pro-Tree Commission, 192
Punitive Expedition, 82–87, 91, 93, 117

radio, 78, 110, 115, 122, 132–34, 137–38, 140–42, 153, 180, 183, 203, 208, 211, 235
Radio Corporation of America (RCA), 134
Railway Lines of Mexico, 173

Ramírez, Margarito, 204
Record, George L., 67–68
Regional Confederation of Mexican Workers (CROM), 205
Regional Federation of Workers and Farmers, 223
Revolutionary Confederation, 56, 58
Reyes, Bernardo, 23–24, 34–37
Richardson, William B., 186
Ricketts, Ed, 191, 247n2
Rivera, Diego, 150, 177–78, 183
Roa, González, 124
roads, 2, 9–10, 43, 62, 115, 130, 142, 180–81, 191–92, 201, 203, 212, 226
Robles, Gonzalo, 168
Rodríguez, Manuel "Manolete," 207
Rojas, Luis Manuel, 88–91
Rolland, Jean François, 2, 5
Rolland, Romain, 95
Rolland Constantine, Jorge M., 231, 236, 238–39
Rolland Garza, Alberto, 22
Rolland Garza, Carmelita, 136
Rolland Garza, Enriqueta, 14, 136, 162
Rolland Garza, Jorge, 39
Rolland Garza, Martha, 22, 162
Rolland Mejía, Victoria, 137, 158, 232–33
Rolland Tolentino, Ana María (Anis), 163, 202, 228–29, 232
Rouaix, Pastor, 58, 61, 63–65, 83, 104, 113, 190
Rowe, Leo S., 91–93
Ruiz, Gregorio, 36
Ruiz Cortines, Adolfo, 218, 223

Saint-Simon, Henri de, 8, 156
Salazar, Luis, 15, 37
Salcido, Jorge, 229
Salinas, León, 109
Salt Workers Cooperative of Marques, 223
Sánchez Fogarty, Federico, 150
Santa Margarita ranch, 200–201, 228, 230–31
scholarly city, 145, 154–55, 157
Scientific and Literary Institute (Hidalgo), 8
Sea of Cortez. *See* Gulf of California
Second on Congress on Town Councils, 130

Segundo Congreso Nacional de Ayuntamientos. *See* Second National Congress on Town Councils
Serrano, Francisco, 155
ship railway, 198, 209–13, 215, 217, 231, 233–34, 240
Sierra, Justo, 4, 12, 67–68, 169, 233
Silva Herzog, Jesús, 221
Simón, Neguib, 206
single tax, 66–67, 87–88, 100, 108, 111
Slaughter, John W., 87
Smiles, Samuel, 72
Smith, R. A. C., 134
social Darwinism, 8
Social-Economic Studies Club, 127–28
socialism: bourgeoisie and, 71, 108; Constitutionalists and, 45–46, 56, 238; democratic, 10; and Francisco Múgica, 181; and Lázaro Cárdenas, 179; and Porfirian-era education, 8, 10, 20; Rolland and, 20, 44–56, 59, 61, 71, 95–96, 108, 193; scientific, 45; wartime, xx
Soto y Gama, Ignacio Díaz, 128, 131
Southeastern Railroad, 173, 182
Spalding, Frederick P., 10
Spanish Civil War, 178–79
Spencer, Herbert, 8
Speyer and Company, 18
stadiums, 148–57
Stampa, Manuel M., 133
Standard Oil, 64
stationary dredge, 198, 213–15, 217, 223–24, 240
Steffens, Lincoln, 67–68, 87–88
Steinbeck, John, xiii–xv, 191
Stiglitz, Joseph, 220
Storey, Moorfield, 90–91
Stridentist, 140–42, 144, 150, 153
Suárez, Eduardo, 187
Suchiate Commission, 202–3

Taft, William H., 33
Tampico, 64
Tannenbaum, Frank, 76
taxes, 106
technocrats, 84
Tehuantepec: challenge to Venustiano Carranza's rule in, 111; environmental degradation in, 242; failure of development in, 139, 221, 226–27, 239, 241; highway across, 212; idea of a canal across, 123, 209; military zone of, 205; people of, 221, 226, 239; ports in, 77, 119, 124–25, 135, 139, 143, 173–74, 186, 197, 200, 212, 221, 224, 226, 241–42; recent projects in, 241–42; resources in, 73, 125; ship railway across, 209–11; strategic location of, 125, 193–95, 199, 208
Tehuantepec Railway, 73, 109, 111, 119, 123–24, 135, 143, 162, 173–74, 186, 193, 197–98, 200, 203–4, 212, 221, 242
Tejada Olivares, Adalberto, 141–42, 145, 150
Ten Tragic Days, 37–38
Tolentino Morales, Rosario, 163, 200–202, 219, 230–32
Tolstoy, Leo, 153, 220
Treviño, Gerónomo, 35
Treviño, Jacinto B., 227
Trotsky, Leon, 177
Tsuru, Kisso, 185

Ugarte, José Gómez, 120
Union of Longshoremen and Laborers, 223
Union of Workers of the Mexican Republic, 223
United States Asphalt Refining Company, 64
Universidad Nacional Autónoma de México (UNAM), 206, 228
Universidad Veracruzana, 238
Urquidi, Francisco, 43, 83
Urquidi, Juan Francisco, 16, 37, 43, 47–49, 83
Urquidi, Manuel, 26, 37, 43, 83
U.S. Congress, 44, 210
U.S. financiers, 240
U.S. General Land Office, 63
U.S. Geological Survey, 63
U.S. Interstate Commerce Commission, 63
U.S. Merchants' Association, 134
U.S. National Guard, 86

Vargas, Getúlio, 179
Vasconcelos, José, 19, 149–50, 169
Vázquez Schiaffino, José, 63
Veracruz Conference (1914–15), 55–59, 62, 65, 88, 118, 143, 181

Villa, Pancho: as Constitutionalist general, 46; fight against the Constitutionalists, 40, 47, 49, 55, 65, 80, 84, 87, 104, 117; Francisco Urquidi's support for, 47–49; as general under Madero, 25; popularity of, 40, 55; and the Punitive Expedition, 84–86; raid on Columbus, 82, 85; as unpredictable, 48, 50; writings against, 49–50
Villagrán García, José, 149
Villar, Lauro, 37
Villarreal, Antonio I., 127, 131–32, 137

Wald, Lillian D., 87
Walsh, Frank P., 89
Wells, H. G., 95
Werf Conrad, 214
Willard, Jess, 62
Wilson, Henry Lane, 32–33, 36–37
Wilson, Woodrow: and the Carrizal incident, 86–87, 90–93, 237; comparison to Constitutionalists, 101; and the invasion of Veracruz, 47; Modesto Rolland's influence on, 84, 101, 237; Modesto Rolland's letter to, 94–95; and the Punitive Expedition, 85–87, 92–93; support for Constitutionalists, 46, 84; supporters of, 44, 84–85; and Victoriano Huerta, 47–48
World Peace Foundation, 89
World War I, 65, 82–83, 106–7, 124, 146, 160, 170
World War II, 182, 194, 205, 208, 220

Wright, Frank Lloyd, 156
Xalapa Stadium, 144–57, 181, 206, 229, 231, 235
Xochimilco waterworks, 16, 21

Yucatán: agrarian reform in, xviii, 11–12, 51, 58, 65–71, 76–77, 81, 110–11, 126–27; concrete construction in, 110; education in, 74, 76–77, 104; feminist movement and congresses in, 71, 73–75, 81; as a frontier peninsula, 62, 114; industrialization in, 71, 73; infrastructure in, 71–73, 77, 81, 108–11; labor and labor policies in, 71–72, 75–77; limited revolutionary violence in, 79, 84; and municipal reform, 90; and the National Agrarian Commission, 126; petroleum and, 73, 77, 100, 108–11; as a positive example of the Revolution, 76, 90, 96; railroads and, 109, 182; removal of Salvador Alvarado from, 105, 110; Salvador Alvarado's military conquest of, 60, 142; and the single tax, 87–88, 100, 108; tropical resources in, 173; and women, 74–75, 81, 222

Zapata, Emiliano, xxii, 25, 33–34, 40, 46, 50, 55, 57, 65, 80, 104; former followers of, 127, 131; local focus of, 55, 238; supporters of 49, 55
Zubarán Capmany, Rafael, 43, 56, 58, 131–32, 137

IN THE MEXICAN EXPERIENCE SERIES

From Idols to Antiquity: Forging the National Museum of Mexico
Miruna Achim

Seen and Heard in Mexico: Children and Revolutionary Cultural Nationalism
Elena Jackson Albarrán

Railroad Radicals in Cold War Mexico: Gender, Class, and Memory
Robert F. Alegre
Foreword by Elena Poniatowska

Mexicans in Revolution, 1910–1946: An Introduction
William H. Beezley and Colin M. MacLachlan

Routes of Compromise: Building Roads and Shaping the Nation in Mexico, 1917–1952
Michael K. Bess

Apostle of Progress: Modesto C. Rolland, Global Progressivism, and the Engineering of Revolutionary Mexico
J. Justin Castro

Radio in Revolution: Wireless Technology and State Power in Mexico, 1897–1938
J. Justin Castro

San Miguel de Allende: Mexicans, Foreigners, and the Making of a World Heritage Site
Lisa Pinley Covert

Celebrating Insurrection: The Commemoration and Representation of the Nineteenth-Century Mexican Pronunciamiento
Edited and with an introduction by Will Fowler

Forceful Negotiations: The Origins of the Pronunciamiento *in Nineteenth-Century Mexico*
Edited and with an introduction by Will Fowler

Independent Mexico: The Pronunciamiento *in the Age of Santa Anna, 1821–1858*
Will Fowler

Malcontents, Rebels, and Pronunciados: *The Politics of Insurrection in Nineteenth-Century Mexico*
Edited and with an introduction by Will Fowler

Working Women, Entrepreneurs, and the Mexican Revolution: The Coffee Culture of Córdoba, Veracruz
Heather Fowler-Salamini

The Heart in the Glass Jar: Love Letters, Bodies, and the Law in Mexico
William E. French

"Muy buenas noches": Mexico, Television, and the Cold War
Celeste González de Bustamante
Foreword by Richard Cole

The Plan de San Diego: Tejano Rebellion, Mexican Intrigue
Charles H. Harris III and Louis R. Sadler

The Inevitable Bandstand: The State Band of Oaxaca and the Politics of Sound
Charles V. Heath

Redeeming the Revolution: The State and Organized Labor in Post-Tlatelolco Mexico
Joseph U. Lenti

Gender and the Negotiation of Daily Life in Mexico, 1750–1856
Sonya Lipsett-Rivera

Mexico's Crucial Century, 1810–1910: An Introduction
Colin M. MacLachlan and William H. Beezley

The Civilizing Machine: A Cultural History of Mexican Railroads, 1876–1910
Michael Matthews

Street Democracy: Vendors, Violence, and Public Space in Late Twentieth-Century Mexico
Sandra C. Mendiola García

The Lawyer of the Church: Bishop Clemente de Jesús Munguía and the Clerical Response to the Liberal Revolution in Mexico
Pablo Mijangos y González

From Angel to Office Worker: Middle-Class Identity and Female Consciousness in Mexico, 1890–1950
Susie S. Porter

¡México, la patria! Propaganda and Production during World War II
Monica A. Rankin

A Revolution Unfinished: The Chegomista Rebellion and the Limits of Revolutionary Democracy in Juchitán, Oaxaca
Colby Ristow

Murder and Counterrevolution in Mexico: The Eyewitness Account of German Ambassador Paul von Hintze, 1912–1914
Edited and with an introduction by Friedrich E. Schuler

Deco Body, Deco City: Female Spectacle and Modernity in Mexico City, 1900–1939
Ageeth Sluis

Pistoleros and Popular Movements: The Politics of State Formation in Postrevolutionary Oaxaca
Benjamin T. Smith

Alcohol and Nationhood in Nineteenth-Century Mexico
Deborah Toner

To order or obtain more information on these or other University of Nebraska Press titles, visit nebraskapress.unl.edu.

www.ingramcontent.com/pod-product-compliance
Lightning Source LLC
Chambersburg PA
CBHW021830220426
43663CB00005B/191